Preface

Kermit Sigmon, author of the MATLAB® Primer, passed away in January 1997. Kermit was a friend, colleague, and fellow avid bicyclist (although I'm a mere 10-mile-a-day commuter) with whom I shared an appreciation for the contribution that MATLAB has made to the mathematics, engineering, and scientific community. MATLAB is a powerful tool, and my hope is that in revising our book for MATLAB 7.0, you will be able to learn how to apply it to solving your own challenging problems in mathematics, science, and engineering.

A team at The MathWorks, Inc. revised the Fifth Edition for MATLAB Version 5 in November of 1997. I carried on Kermit's work by creating the Sixth Edition of this book for MATLAB 6.1 in October 2001, and now this Seventh Edition for MATLAB Version 7.0.

This edition highlights the many new features of MATLAB 7.0, and includes new chapters on features that were in prior versions of MATLAB but not in prior editions of this book. New or revised topics in this edition include:

- calling Java from MATLAB, and using Java objects inside the MATLAB workspace
- many more graphics examples, including the seashell on the cover of the book
- cell publishing for reports in HTML, LaTeX, Microsoft Word, and Microsoft Powerpoint
- powerful suite of code development tools (such as the M-Lint code checker, the file dependency and comparison reports, and a profile coverage report)

- volume and vector visualization
- calling Fortran code from MATLAB
- parametric curves and surfaces, and polar plots of symbolic functions
- polynomials, interpolation, and numeric integration
- solving non-linear equations with `fzero`
- solving ordinary differential equations with `ode45`
- the revised MATLAB Desktop
- short-circuit logical operators
- integers and single precision floating-point
- more details on the colon operator
- `linsolve`, for solving specific linear systems
- the new block comment syntax
- function handles (@), which are now simpler to use
- anonymous functions
- `image`, and a pretty Mandelbrot set example
- the new 4-output sparse `lu`
- abstract symbolic functions
- nicely-formatted tables using `fprintf`
- a revised list of all primary functions and operators in MATLAB.

I would like to thank Penny Anderson at The MathWorks, Inc. for her detailed review of this book.

Tim Davis
Associate Professor, Department of Computer and Information Science and Engineering, University of Florida, http://www.cise.ufl.edu/research/sparse

Introduction

MATLAB®, developed by The MathWorks, Inc., integrates computation, visualization, and programming in a flexible, open environment. It offers engineers, scientists, and mathematicians an intuitive language for expressing problems and their solutions mathematically and graphically. Complex numeric and symbolic problems can be solved in a fraction of the time required with a programming language such as C, Fortran, or Java.

How to use this book: The purpose of this Primer is to help you begin to use MATLAB. It is not intended to be a substitute for the online help facility or the MATLAB documentation (such as *Getting Started with MATLAB*, available in printed form and online). The Primer can best be used hands-on. You are encouraged to work at the computer as you read the Primer and freely experiment with the examples. This Primer, along with the online help facility, usually suffices for students in a class requiring the use of MATLAB.

Start with the examples at the beginning of each chapter. In this way, you will create all of the matrices and M-files used in the examples. Some examples depend on code you write in previous chapters.

Larger examples (M-files and MEX-files) are on the web at http://www.cise.ufl.edu/research/sparse/MATLAB and http://www.crcpress.com.

Pull-down menu selections are described using the following style. Selecting the `Desktop` menu, and then the `Desktop Layout` submenu, and then the `Default` menu item is written as Desktop ▶ Desktop Layout ▶ Default.

You should liberally use the online help facility for more detailed information. Pressing the F1 key or selecting Help ▶ MATLAB Help brings up the Help window. You can also type help or doc in the Command window. See Sections 2.1 or 22.26 for more information on how to use the online help.

How to obtain MATLAB: Version 7.0 (Release 14) of MATLAB is available for Microsoft Windows (XP, 2000, and NT 4.0), Unix (Linux, Solaris 2.8 and 2.9, and HP-UX 11 or 11i), and the Macintosh (OS X 10.3.2 Panther). A Student Version is available for all but Solaris and HP-UX; it includes MATLAB, Simulink, and key functions of the Symbolic Math Toolbox. Everything discussed in this book can be done in the Student Version of MATLAB, with the exception of advanced features of the Symbolic Math Toolbox discussed in Section 16.13.

MATLAB, Simulink, Handle Graphics, StateFlow, and Real-Time Workshop are registered trademarks of The MathWorks, Inc. TargetBox is a trademark of The MathWorks, Inc. For more information on MATLAB, contact:

The MathWorks, Inc.
3 Apple Hill Drive
Natick, MA, 01760-2098 USA
Phone: 508–647–7000
Fax: 508–647–7101
Web: http://www.mathworks.com

Table of Contents

17. Polynomials, Interpolation, and Integration.. 118

18. Solving Equations...................................... 122

19. Displaying Results..................................... 128
20. Cell Publishing.. 132
21. Code Development Tools........................... 133

1. Accessing MATLAB

On Unix systems you can enter MATLAB with the system command `matlab` and exit MATLAB with the MATLAB command `quit` or `exit`. In Microsoft Windows and the Macintosh, just double-click on the MATLAB icon:

MATLAB 7.0

2. The MATLAB Desktop

MATLAB has an extensive graphical user interface. When MATLAB starts, the MATLAB window will appear, with several subwindows and menu bars.

All of MATLAB's windows in the default desktop are docked, which means that they are tiled on the main MATLAB window. You can undock a window by selecting the menu item Desktop ▶ ⫟ Undock or by clicking its undock button:

⫟

Dock it with Desktop ▶ ⫟ Dock... or the dock button:

⫟

Close a window by clicking its close button:

✕

Reshape the window tiling by clicking on and dragging the window edges.

The menu bar at the top of the MATLAB window contains a set of buttons and pull-down menus for working with M-files, windows, preferences and other settings, web resources for MATLAB, and online MATLAB help. If a window is docked and selected, its menu bar appears at the top of the MATLAB window.

If you prefer a simpler font than the default one, select File ▸ Preferences, and click on ⊞ Fonts. Select Lucida Console (on a PC) or DialogInput (on Unix) in place of the default Monospaced font, and click OK.

2.1 Help window

This window is the most useful window for beginning MATLAB users, and MATLAB experts continue to use it heavily. Select Help ▸ MATLAB Help or type doc. The Help window has most of the features you would see in any web browser (clickable links, a back button, and a search engine, for example). The Help Navigator on the left shows where you are in the MATLAB online documentation. Online Help sections are referred to as Help: MATLAB: Getting Started: Introduction, for example. Click on the ⊞ beside MATLAB in the Help Navigator, and you will see the MATLAB Roadmap (or Help: MATLAB for short). Printable versions of the documentation are available under this category (see Help: MATLAB: Printable Documentation (PDF)).

You can also use the help command, typed in the Command window. For example, the command help eig will give information about the eigenvalue function eig. See the list of functions in Chapter 22 for a brief summary of help for a function. doc is similar, except that it displays information in the Help Browser. You can also preview some of the features of MATLAB by first entering the command **demo** or by selecting **Help ▶ Demos**, and then selecting from the options offered.

2.2 Start button

The Start button in the bottom left corner of the MATLAB Desktop allows you to start up demos, tools, and other windows not present when you start MATLAB. Try **Start**: **MATLAB**: **Demos** and run one of the demos from the MATLAB Demo window.

2.3 Command window

MATLAB expressions and statements are evaluated as you type them in the Command window, and results of the computation are displayed there too. Expressions and statements are also used in M-files (more on this in Chapter 7). They are usually of the form:

> *variable = expression*

or simply:

> *expression*

Expressions are usually composed from operators, functions, and variable names. Evaluation of the expression produces a matrix (or other data type), which is then displayed on the screen or assigned to a variable for future use. If the variable name and = sign are omitted, a variable **ans** (for answer) is automatically created to which the result is assigned.

A statement is normally terminated at the end of the line. However, a statement can be continued to the next line with three periods (. . .) at the end of the line. Several statements can be placed on a single line separated by commas or semicolons. If the last character of a statement is a semicolon, display of the result is suppressed, but the assignment is carried out. This is essential in suppressing unwanted display of intermediate results.

Click on the Workspace tab to bring up the Workspace window (it starts out underneath the Current Directory window in the default layout) so you can see a list of the variables you create, and type this command in the Command window:

```
A = [1 2 3 ; 4 5 6 ; -1 7 9]
```

or this one:

```
A = [
1 2 3
4 5 6
-1 7 9]
```

in the Command window. Either one creates the obvious 3-by-3 matrix and assigns it to a variable A. Try it. You will see the array A in your Workspace window. MATLAB is case-sensitive in the names of commands, functions, and variables, so A and a are two different variables. A comma or blank separates the elements within a row of a matrix (sometimes a comma is necessary to split the expressions, because a blank can be ambiguous). A semicolon ends a row. When listing a number in exponential form (e.g., 2.34e-9), blank spaces must be avoided in the middle (before the e, for example). Matrices can also be constructed from other matrices. If A is the 3-by-3 matrix shown above, then:

```
C = [A, A' ; [12 13 14], zeros(1,3)]
```

creates a 4-by-6 matrix. Try it to see what C is. The
quote mark in A' means the transpose of A. Be sure to
use the correct single quote mark (just to the left of the
enter or return key on most keyboards). Since a blank
separates elements in a row, parentheses are sometimes
needed around expressions if they would otherwise be
ambiguous. See Section 5.1 for the zeros function.

When you typed the last two commands, the matrices A
and C were created and displayed in the Workspace
window.

You can save the Command window dialog with the
diary command:

    diary *filename*

This causes what appears subsequently in the Command
window to be written to the named file (if the *filename*
is omitted, it is written to a default file named diary)
until you type the command diary off; the command
diary on causes writing to the file to resume. When
finished, you can edit the file as desired and print it out.
For hard copy of graphics, see Section 12.10.

The command line in MATLAB can be easily edited in
the Command window. The cursor can be positioned
with the left and right arrows and the Backspace (or
Delete) key used to delete the character to the left of the
cursor.

A convenient feature is use of the up and down arrows to
scroll through the stack of previous commands. You can,

therefore, recall a previous command line, edit it, and execute the revised line. Try this by first modifying the matrix A by adding one to each of its elements:

```
A = A + 1
```

You can change C to reflect this change in A by retyping the lengthy command C = ... above, but it is easier to hit the up arrow key until you see the command you want, and then hit enter.

You can clear the Command window with the `clc` command or with Edit ▶ Clear Command Window.

The format of the displayed output can be controlled by the following commands:

format short	fixed point, 5 digits
format long	fixed point, 15 digits
format short e	scientific notation, 5 digits
format long e	scientific notation, 15 digits
format short g	fixed or floating-point, 5 digits
format long g	fixed or floating-point, 15 digits
format hex	hexadecimal format
format '+'	+, -, and blank
format bank	dollars and cents
format rat	approximate integer ratio

format short is the default. Once invoked, the chosen format remains in effect until changed. These commands only modify the display, not the precision of the number or its computation. Most numeric computations in MATLAB are done in double precision, which has about 16 digits of accuracy.

The command `format compact` suppresses most blank lines, allowing more information to be placed on the screen or page. The command `format loose` returns to the non-compact format. These two commands are independent of the other format commands.

You can pause the output in the Command window with the `more on` command. Type `more off` to turn this feature off.

2.4 Workspace window

The Workspace window lists variables that you have either entered or computed in your MATLAB session.

There are many fundamental data types (or classes) in MATLAB, each one a multidimensional array. The classes that we will concern ourselves with most are rectangular numerical arrays with possibly complex entries, and possibly sparse. An array of this type is called a matrix. A matrix with only one row or one column is called a vector (row vectors and column vectors behave differently; they are more than mere one-dimensional arrays). A 1-by-1 matrix is called a scalar.

Arrays can be introduced into MATLAB in several different ways. They can be entered as an explicit list of elements (as you did for matrix A), generated by statements and functions (as you did for matrix C), created in a file with your favorite text editor, or loaded from external data files or applications (see `Help: MATLAB: Getting Started: Manipulating Matrices`). You can also write your own functions (M-files, mexFunctions in C or Fortran, or Java) that create and operate on matrices. All the matrices and other variables that you create, except those internal to M-files, are shown in your Workspace window.

The command who (or whos) lists the variables currently in the workspace. Try typing whos; you should see a list of variables including A and C, with their type and size. A variable or function can be cleared from the workspace with the command `clear variablename` or by right-clicking the variable in the Workspace editor and selecting `Delete`. The command `clear` alone clears all variables from the workspace.

When you log out or exit MATLAB, all variables are lost. However, invoking the command `save` before exiting causes all variables to be written to a machine-readable file named `matlab.mat` in the current working directory. When you later reenter MATLAB, the command `load` will restore the workspace to its former state. Commands `save` and `load` take file names and variable names as optional arguments (type `doc save` and `doc load`). Try typing the commands `save`, `clear`, and then `load`, and watch what happens in the Workspace window after each command.

2.5 Command History window

This window lists the commands typed in so far. You can re-execute a command from this window by double-clicking or dragging the command into the Command window. Try double-clicking on the command:

```
A = A + 1
```

shown in your Command History window. For more
options, select and right-click on a line of the Command
window.

2.6 Array Editor window

Once an array exists, it can be modified with the Array
Editor, which acts like a spreadsheet for matrices. Go to
the Workspace window and double-click on the matrix C.
Click on an entry in C and change it, and try changing the
size of C. Go back to the Command window and type:

```
C
```

and you will see your new array C. You can also edit the
matrix C by typing the command openvar('C').

2.7 Current Directory window

Your current directory is where MATLAB looks for your
M-files, and for workspace (.mat) files that you load
and save. You can also load and save matrices as ASCII
files and edit them with your favorite text editor. The file
should consist of a rectangular array of just the numeric
matrix entries. Use a text editor to create a file in your
current directory called mymatrix.txt (or type edit
mymatrix.txt) that contains these 2 lines:

```
22 67
12 33
```

Type the command load mymatrix.txt, and the file
will be loaded from the current directory to the variable
mymatrix. The file extension (.txt in this example)
can be anything except .mat.

You can use the menus and buttons in the Current Directory window to peruse your files, or you can use commands typed in the Command window. The command pwd returns the name of the current directory, and cd will change the current directory. The command dir lists the contents of the working directory, whereas the command what lists only the MATLAB-specific files in the directory, grouped by file type. The MATLAB commands delete and type can be used to delete a file and display a file in the Command window, respectively.

The Current Directory window includes a suite of useful code development tools, described in Chapter 21.

3. Matrices and Matrix Operations

You have now seen most of MATLAB's windows and what they can do. Now take a look at how you can use MATLAB to work on matrices and other data types.

3.1 Referencing individual entries

Individual matrix and vector entries can be referenced with indices inside parentheses. For example, A(2,3) denotes the entry in the second row, third column of matrix A. Try:

```
A = [1 2 3 ; 4 5 6 ; -1 7 9]
A(2,3)
```

Next, create a column vector, x, with:

```
x = [3 2 1]'
```

or equivalently:

```
x = [3 ; 2 ; 1]
```

With this vector, x(3) denotes the third coordinate of vector x, with a value of 1. Higher dimensional arrays are similarly indexed. An array accepts only positive integers as indices.

An array with two or more dimensions can be indexed as if it were a one-dimensional vector. If A is m-by-n, then A(i,j) is the same as A(i+(j-1)*m). This feature is most often used with the find function (see Section 5.6).

3.2 Matrix operators

The following matrix operators are available in MATLAB:

+   addition or unary plus
−   subtraction or negation
*   multiplication
^   power
'   transpose (real) or conjugate transpose (complex)
.'   transpose (real or complex)
\   left division (*backslash* or mldivide)
/   right division (*slash* or mrdivide)

These matrix operators apply, of course, to scalars (1-by-1 matrices) as well. If the sizes of the matrices are incompatible for the matrix operation, an error message will result, except in the case of scalar-matrix operations (for addition, subtraction, division, and multiplication, in which case each entry of the matrix is operated on by the scalar, as in A=A+1). Not all scalar-matrix operations are valid. For example, magic(3)/pi is valid but pi/magic(3) is not. Also try the commands:

```
A^2
A*x
```

If x and y are both column vectors, then x'*y is their inner (or dot) product, and x*y' is their outer (or cross) product. Try these commands:

```
y = [1 2 3]'
x'*y
x*y'
```

3.3 Matrix division (slash and backslash)

The matrix "division" operations deserve special comment. If A is an invertible square matrix and b is a compatible column vector, or respectively a compatible row vector, then x=A\b is the solution of A*x=b, and x=b/A is the solution of x*A=b. These are also called the backslash (\) and slash operators (/); they are also referred to as the mldivide and mrdivide functions.

If A is square and non-singular, then A\b and b/A are mathematically the same as inv(A)*b and b*inv(A), respectively, where inv(A) computes the inverse of A. The left and right division operators are more accurate and efficient. In left division, if A is square, then it is factorized (if necessary), and these factors are used to solve A*x=b. If A is not square, the under- or over-determined system is solved in the least squares sense. Right division is defined in terms of left division by b/A = (A'\b')'. Try this:

```
A = [1 2 ; 3 4]
b = [4 10]'
x = A\b
```

The solution to A*x=b is the column vector x=[2;1].

Backslash is a very powerful general-purpose method for solving linear systems. Depending on the matrix, it selects forward or back substitution for triangular matrices (or permuted triangular matrices), Cholesky factorization for symmetric matrices, LU factorization for square matrices, or QR factorization for rectangular matrices. It has a special solver for Hessenberg matrices. It can also exploit sparsity, with either sparse versions of the above list, or special-case solvers when the sparse matrix is diagonal, tridiagonal, or banded. It selects the best method automatically (sometimes trying one method and then another if the first method fails). This can be overkill if you already know what kind of matrix you have. It can be much faster to use the `linsolve` function described in Section 5.5.

3.4 Entry-wise operators

Matrix addition and subtraction already operate entry-wise, but the other matrix operations do not. These other operators (*, ∧, \, and /) can be made to operate entry-wise by preceding them by a period. For example, either:

```
[1 2 3 4] .* [1 2 3 4]
[1 2 3 4] .∧ 2
```

will yield [1 4 9 16]. Try it. This is particularly useful when using MATLAB graphics.

Also compare A∧2 with A.∧2.

3.5 Relational operators

The relational operators in MATLAB are:

<	less than
>	greater than
<=	less than or equal
>=	greater than or equal
==	equal
~=	not equal

They all operate entry-wise. Note that = is used in an assignment statement whereas == is a relational operator. Relational operators may be connected by logical operators:

&	and
|	or
~	not
&&	short-circuit and
||	short-circuit or

The result of a relational operator is of type `logical`, and is either `true` (one) or `false` (zero). Thus, ~0 is 1, ~3 is 0, and 4 & 5 is 1, for example. When applied to scalars, the result is a scalar. Try entering 3 < 5, 3 > 5, 3 == 5, and 3 == 3. When applied to matrices of the same size, the result is a matrix of ones and zeros giving the value of the expression between corresponding entries. You can also compare elements of a matrix with a scalar. Try:

```
A = [1 2 ; 3 4]
A >= 2
B = [1 3 ; 4 2]
A < B
```

The short-circuit operator && acts just like its non-short-circuited counterpart (&), except that it evaluates its left expression first, and does not evaluate the right expression if the first expression is `false`. This is useful for partially-defined functions. Suppose f(x) returns a logical value but generates an error if x is zero. The expression (x~=0) && f(x) returns `false` if x is zero, without calling f(x) at all. The short-circuit or (||) acts similarly. It does not evaluate the right expression if the left is `true`. Both && and || require their operands to be scalar and convertible to logical, while & and | can operate on arrays.

3.6 Complex numbers

MATLAB allows complex numbers in most of its operations and functions. Three convenient ways to enter complex matrices are:

```
B = [1 2 ; 3 4] + i*[5 6 ; 7 8]
B = [1+5i, 2+6i ; 3+7i, 4+8i]
B = complex([1 2 ; 3 4], [5 6 ; 7 8])
```

Either i or j may be used as the imaginary unit. If, however, you use i and j as variables and overwrite their values, you may generate a new imaginary unit with, say, ii=sqrt(-1). You can also use 1i or 1j, which cannot be reassigned and are always equal to the imaginary unit. Thus,

```
B = [1 2 ; 3 4] + 1i*[5 6 ; 7 8]
```

generates the same matrix B, even if i has been reassigned. See Section 8.2 for how to find out if i has been reassigned.

3.7 Strings

Enclosing text in single quotes forms strings with the char data type:

```
S = 'I love MATLAB'
```

To include a single quote inside a string, use two of them together, as in:

```
S = 'Green''s function'
```

Strings, numeric matrices, and other data types can be displayed with the function disp. Try disp(S) and disp(B).

3.8 Other data types

MATLAB supports many other data types, including logical variables, integers of various sizes, single-precision floating-point variables, sparse matrices, multidimensional arrays, cell arrays, and structures.

The default data type is double, a 64-bit IEEE floating-point number. The single type is a 32-bit IEEE floating-point number which should be used only if you are desperate for memory. A double can represent integers in the range -2^{53} to 2^{53} without any roundoff error, and a double holding an integer value is typically used for loop and array indices. An integer value stored as a double is nicknamed a *flint*. Integer types are only needed in special cases such as signal processing, image processing, encryption, and bit string manipulation. Integers come in signed and unsigned flavors, and in sizes of 8, 16, 32, and 64 bits. Integer arithmetic is not modular, but saturates on overflow. If you want a warning to be generated when integers overflow, use `intwarning on`. See `doc int8` and `doc single` for more information.

A sparse matrix is not actually its own data type, but an attribute of the `double` and `logical` matrix types. Sparse matrices are stored in a special way that does not require space for zero entries. MATLAB has efficient methods of operating on sparse matrices. Type `doc sparse`, and `doc full`, look in `Help: MATLAB: Mathematics: Sparse Matrices`, or see Chapter 15. Sparse matrices are allowed as arguments for most, but not all, MATLAB operators and functions where a normal matrix is allowed.

`D=zeros(3,5,4,2)` creates a 4-dimensional array of size 3-by-5-by-4-by-2. Multidimensional arrays may also be built up using `cat` (short for concatenation).

Cell arrays are collections of other arrays or variables of varying types and are formed using curly braces. For example,

```
c = {[3 2 1] ,'I love MATLAB'}
```

creates a cell array. The expression `c{1}` is a row vector of length 3, while `c{2}` is a string.

A `struct` is variable with one or more parts, each of which has its own type. Try, for example,

```
x.particle = 'electron'
x.position = [2 0 3]
x.spin = 'up'
```

17

The variable x describes an object with several characteristics, each with its own type.

You may create additional data objects and classes using overloading (see `help class` or `doc class`).

4. Submatrices and Colon Notation

Vectors and submatrices are often used in MATLAB to achieve fairly complex data manipulation effects. Colon notation (which is used to both generate vectors and reference submatrices) and subscripting by integral vectors are keys to efficient manipulation of these objects. Creative use of these features minimizes the use of loops (which can slow MATLAB) and makes code simple and readable. Special effort should be made to become familiar with them.

4.1 Generating vectors

The expression `1:5` is the row vector `[1 2 3 4 5]`. The numbers need not be integers, and the increment need not be one. For example, `0:0.2:1` gives `[0 0.2 0.4 0.6 0.8 1]` with an increment of `0.2` and `5:-1:1` gives `[5 4 3 2 1]` with an increment of `-1`. These vectors are commonly used in `for` loops, described in Section 6.1. Be careful how you mix the colon operator with other operators. Compare `1:5-3` with `(1:5)-3`.

In general, the expression `lo:hi` is the sequence `[lo, lo+1, lo+2, …, hi]` except that the last term in the sequence is always less than or equal to `hi` if either one are not integers. Thus, `1:4.9` is `[1 2 3 4]` and `1:5.1` is `[1 2 3 4 5]`. The sequence is empty if `lo > hi`.

If an increment is provided, as in `lo:inc:hi`, then the sequence is `[lo, lo+inc, lo+2*inc, …, lo+m*inc]` where `m=fix((hi-lo)/inc)` and `fix` is a function that rounds a real number towards zero. The length of the sequence is m+1, and the sequence is empty if m<0. Thus, the sequence `5:-1:1` has m=4 and is of length 5, but `5:1:1` has m=-4 and is thus empty. The default increment is 1.

If you want specific control over how many terms are in the sequence, use `linspace` instead of the colon operator. The expression `linspace(lo,hi)` is identical to `lo:inc:hi`, except that `inc` is chosen so that the vector always has exactly 100 entries (even if `lo` and `hi` are equal). The last entry in the sequence is always `hi`. To generate a sequence with n terms instead of the default of 100, use `linspace(lo,hi,n)`. Compare `linspace(1,5.1,5)` with `1:5.1`.

4.2 Accessing submatrices

Colon notation can be used to access submatrices of a matrix. To try this out, first type the two commands:

```
A = rand(6,6)
B = rand(6,4)
```

which generate a random 6-by-6 matrix A and a random 6-by-4 matrix B.

`A(1:4,3)` is the column vector consisting of the first four entries of the third column of A.

A colon by itself denotes an entire row or column:
`A(:,3)` is the third column of A, and `A(1:4,:)` is the first four rows.

Arbitrary integral vectors can be used as subscripts: A(:, [2 4]) contains as columns, columns 2 and 4 of A. Such subscripting can be used on both sides of an assignment statement:

```
A(:,[2 4 5]) = B(:,1:3)
```

replaces columns 2,4,5 of A with the first three columns of B. Try it. Note that the entire altered matrix A is displayed and assigned. Columns 2 and 4 of A can be multiplied on the right by the matrix [1 2 ; 3 4]:

```
A(:,[2 4]) = A(:,[2 4]) * [1 2 ; 3 4]
```

Once again, the entire altered matrix is displayed and assigned. Submatrix operations are a convenient way to perform many useful computations. For example, a Givens rotation of rows 3 and 5 of the matrix A to zero out the A(3,1) entry can be written as:

```
a = A(5,1)
b = A(3,1)
G = [a b ; -b a] / norm([a b])
A([5 3], :) = G * A([5 3], :)
```

(assuming norm([a b]) is not zero). You can also assign a scalar to all entries of a submatrix. Try:

```
A(:, [2 4]) = 99
```

You can delete rows or columns of a matrix by assigning the empty matrix ([]) to them. Try:

```
A(:, [2 4]) = []
```

In an array index expression, end denotes the index of the last element. Try:

```
x = rand(1,5)
x = x(end:-1:1)
```

To appreciate the usefulness of these features, compare
these MATLAB statements with a C, Fortran, or Java
routine to do the same operation.

5. MATLAB Functions

MATLAB has a wide assortment of built-in functions.
You have already seen some of them, such as zeros,
rand, and inv. This section describes the more common
matrix manipulation functions. For a more complete list,
see Chapter 22, or Help: MATLAB: Functions --
Categorical List.

5.1 Constructing matrices

Convenient matrix building functions include:

eye	identity matrix
zeros	matrix of zeros
ones	matrix of ones
diag	create or extract diagonals
triu	upper triangular part of a matrix
tril	lower triangular part of a matrix
rand	randomly generated matrix
hilb	Hilbert matrix
magic	magic square
toeplitz	Toeplitz matrix
gallery	a wide range of interesting matrices

The command rand(n) creates an n-by-n matrix with
randomly generated entries distributed uniformly between
0 and 1 while rand(m,n) creates an m-by-n matrix (m
and n are non-negative integers). Try:

```
A = rand(3)
```

rand('state',0) resets the random number generator.
zeros(m,n) produces an m-by-n matrix of zeros, and
zeros(n) produces an n-by-n one. If A is a matrix, then
zeros(size(A)) produces a matrix of zeros having the
same size as A. If x is a vector, diag(x) is the diagonal
matrix with x down the diagonal; if A is a matrix, then
diag(A) is a vector consisting of the diagonal of A. Try:

```
x = 1:3
diag(x)
diag(A)
diag(diag(A))
```

Matrices can be built from blocks. Try creating this 5-by-
5 matrix.

```
B = [A zeros(3,2) ; pi*ones(2,3), eye(2)]
```

magic(n) creates an n-by-n matrix that is a magic
square (rows, columns, and diagonals have common
sum); hilb(n) creates the n-by-n Hilbert matrix, a very
ill-conditioned matrix. Matrices can also be generated
with a for loop (see Section 6.1). triu and tril extract
upper and lower triangular parts of a matrix. Try:

```
triu(A)
triu(A) == A
```

The gallery function can generate a matrix from any
one of over 60 different matrix classes. Many have
interesting eigenvalue or singular value properties,
provide interesting counter-examples, or are difficult
matrices for various linear algebraic methods. The
Rosser matrix challenges many eigenvalue solvers:

```
A = gallery('rosser')
eig(A)
eigs(A)
```

The Parter matrix has many singular values close to π:

```
A = gallery('parter', 6)
svd(A)
```

The eig, eigs, and svd functions are discussed below.

5.2 Scalar functions

Certain MATLAB functions operate essentially on scalars
but operate entry-wise when applied to a vector or matrix.
Some of the most common such functions are:

```
abs    ceil   floor   rem      sqrt
acos   cos    log     round    tan
asin   exp    log10   sign
atan   fix    mod     sin
```

The following statements will generate a sine table:

```
x = (0:0.1:2)'
y = sin(x)
[x y]
```

Note that because sin operates entry-wise, it produces a
vector y from the vector x.

5.3 Vector functions and data analysis

Other MATLAB functions operate essentially on a vector
(row or column) but act on an m-by-n matrix (m > 2) in a
column-by-column fashion to produce a row vector
containing the results of their application to each column.
Row-by-row action can be obtained by using the
transpose (mean(A')', for example) or by specifying the

dimension along which to operate (`mean(A,2)`, for example). Most of these functions perform basic statistical computations (`std` computes the standard deviation and `prod` computes the product of the elements in the vector, for example). The primary functions are:

```
max   sum   median  any  sort  var
min   prod  mean    all  std
```

The maximum entry in a matrix A is given by `max(max(A))` rather than `max(A)`. Try it. The `any` and `all` functions are discussed in Section 6.6.

5.4 Matrix functions

Much of MATLAB's power comes from its matrix functions. Here is a partial list of the most common ones:

`eig`	eigenvalues and eigenvectors
`eigs`	like `eig`, for large sparse matrices
`chol`	Cholesky factorization
`svd`	singular value decomposition
`svds`	like `svd`, for large sparse matrices
`inv`	inverse
`lu`	LU factorization
`qr`	QR factorization
`hess`	Hessenberg form
`schur`	Schur decomposition
`rref`	reduced row echelon form
`expm`	matrix exponential
`sqrtm`	matrix square root
`poly`	characteristic polynomial
`det`	determinant
`size`	size of an array
`length`	length of a vector

norm	1-norm, 2-norm, Frobenius-norm, ∞-norm
normest	2-norm estimate
cond	condition number in the 2-norm
condest	condition number estimate
rank	rank
kron	Kronecker tensor product
find	find indices of nonzero entries
linsolve	solve a special linear system

MATLAB functions may have single or multiple output arguments. Square brackets are used to the left of the equal sign to list the outputs. For example,

```
y = eig(A)
```

produces a column vector containing the eigenvalues of A, whereas:

```
[V, D] = eig(A)
```

produces a matrix V whose columns are the eigenvectors of A and a diagonal matrix D with the eigenvalues of A on its diagonal. Try it. The matrix A*V-V*D will have small entries.

5.5 The linsolve function

The matrix divide operators (\ or /) are usually enough for solving linear systems. They look at the matrix and try to pick the best method. The linsolve function acts like \, except that you can tell it about your matrix. Try:

```
A = [1 2 ; 3 4]
b = [4 10]'
A\b
linsolve(A,b)
```

In both cases, you get solution x=[2;1] to the linear system A*x=b.

If A is symmetric and positive definite, one explicit solution method is to perform a Cholesky factorization, followed by two solves with triangular matrices. Try:

```
C = [2 1 ; 1 2]
x = C\b
```

Here is an equivalent method:

```
R = chol(C)
y = R'\b
x = R\y
```

The matrix R is upper triangular, but MATLAB explicitly transposes R and then determines for itself that R' is lower triangular. You can save MATLAB some work by using linsolve with an optional third argument, opts. Try this:

```
opts.UT = true
opts.TRANSA = true
y = linsolve(R,b,opts)
```

which gives the same answer as y=R'\b. The difference in run time can be high for large matrices (see Chapter 10 for more details). The fields for opts are UT (upper triangular), LT (lower triangular), UHESS (upper Hessenberg), SYM (symmetric), POSDEF (positive definite), RECT (rectangular), and TRANSA (whether to solve A*x=b or A'*x=b). All opts fields are either true or false. Not all combinations are supported (type doc linsolve for a list). linsolve does not work on sparse matrices.

5.6 The find function

The find function is unlike the other matrix and vector functions. find(x), where x is a vector, returns an array of indices of nonzero entries in x. This is often used in conjunction with relational operators. Suppose you want a vector y that consists of all the values in x greater than 1. Try:

```
x = 2*rand(1,5)
y = x(find(x > 1))
```

With three output arguments, you get more information:

```
A = rand(3)
[i,j,x] = find(A)
```

returns three vectors, with one entry in i, j, and x for each nonzero in A (row index, column index, and numerical value, respectively). With this matrix A, try:

```
[i,j,x] = find(A > .5)
[i j x]
```

and you will see a list of pairs of row and column indices where A is greater than .5. However, x is a vector of values from the matrix expression A > .5, not from the matrix A. Getting the values of A that are larger than .5 without a loop requires one-dimensional array indexing:

```
k = find(A > .5)
A(k)
A(k) = A(k) + 99
```

Section 6.1 shows the loop-based version of this code.

Here is a more complex example. A square matrix A is diagonally dominant if

$$|a_{ii}| > \sum_{j \neq i} |a_{ij}| \qquad \text{for each row } i.$$

First, enter a matrix that is not diagonally dominant. Try:

```
A = [
-1  2   3 -4
 0  2  -1  0
 1  2   9  1
-3  4   1  1]
```

These statements compute a vector i containing indices of rows that violate diagonal dominance (rows 1 and 4 for this matrix A).

```
d = diag(A)
a = abs(d)
f = sum(abs(A), 2) - a
i = find(f >= a)
```

Next, modify the diagonal entries to make the matrix just barely diagonally dominant, while still preserving the sign of the diagonal:

```
[m n] = size(A)
k = i + (i-1)*m
tol = 100 * eps
s = 2 * (d(i) >= 0) - 1
A(k) = (1+tol) * s .* max(f(i), tol)
```

The variable eps (epsilon) gives the smallest value such that $1+eps > 1$, about 10^{-16} on most computers. It is useful in specifying tolerances for convergence of iterative processes and in problems like this one. The odd-looking statement that computes s is nearly the same as s=sign(d(i)), except that here we want s to be one when d(i) is zero. We will come back to this diagonal dominance problem later on.

6. Control Flow Statements

In their basic forms, these MATLAB flow control statements operate like those in most computer languages. Indenting the statements of a loop or conditional statement is optional, but it helps readability to follow a standard convention.

6.1 The for loop

This loop:

```
n = 10
x = []
for i = 1:n
    x = [x, i^2]
end
```

produces a vector of length 10, and

```
n = 10
x = []
for i = n:-1:1
    x = [i^2, x]
end
```

produces the same vector. Try them. The vector x grows in size at each iteration. Note that a matrix may be empty (such as x=[]). The statements:

```
m = 6
n = 4
for i = 1:m
    for j = 1:n
```

```
                   H(i,j) = 1/(i+j-1) ;
        end
    end
    H
```

produce and display in the Command window the 6-by-4
Hilbert matrix. The last H displays the final result. The
semicolon on the inner statement is essential to suppress
the display of unwanted intermediate results. If you leave
off the semicolon, you will see that H grows in size as the
computation proceeds. This can be slow if m and n are
large. It is more efficient to preallocate the matrix H with
the statement H=zeros(m,n) before computing it. Type
the command doc hilb and type hilb to see a more
efficient way to produce a square Hilbert matrix.

Here is the counterpart of the one-dimensional indexing
exercise from Section 5.6. It adds 99 to each entry of the
matrix that is larger than .5, using two for loops instead
of a single find. This method is slower:

```
    A = rand(3)
    [m n] = size(A) ;
    for j = 1:n
        for i = 1:m
            if (A(i,j) > .5)
                A(i,j) = A(i,j) + 99 ;
            end
        end
    end
    A
```

The for statement permits any matrix expression to be
used instead of 1:n. The index variable consecutively
assumes the value of each column of the expression. For
example,

```
s = 0 ;
for c = H
    s = s + sum(c) ;
end
```

computes the sum of all entries of the matrix H by adding
its column sums (of course, sum(sum(H)) does it more
efficiently; see Section 5.3). Each iteration of the for
loop assigns a successive column of H to the variable c.
In fact, since 1:n = [1 2 3 ... n], this column-by-
column assignment is what occurs with for i = 1:n.

6.2 The while loop

The general form of a while loop is:

```
while expression
    statements
end
```

The *statements* will be repeatedly executed as long as
the *expression* remains true. For example, for a given
number a, the following computes and displays the
smallest nonnegative integer n such that $2^n > a$:

```
a = 1e9
n = 0
while 2^n <= a
    n = n + 1 ;
end
n
```

Note that you can compute the same value n more
efficiently by using the log2 function:

```
[f,n] = log2(a)
```

You can terminate a for or while loop with the break
statement and skip to the next iteration with the

`continue` statement. Here is an example for both. It prints the odd integers from 1 to 7 by skipping over the even iterations and then terminates the loop when `i` is 7.

```
for i = 1:10
    if (mod(i,2) == 0)
        continue
    end
    i
    if (i == 7)
        break
    end
end
```

6.3 The if statement

The general form of a simple `if` statement is:

```
if expression
    statements
end
```

The *statements* will be executed only if the *expression* is true. Multiple conditions also possible:

```
for n = -2:5
    if n < 0
        parity = 0 ;
    elseif rem(n,2) == 0
        parity = 2 ;
    else
        parity = 1 ;
    end
    disp([n parity])
end
```

The `else` and `elseif` are optional. If the `else` part is used, it must come last.

6.4 The switch statement

The `switch` statement is just like the `if` statement. If you have one expression that you want to compare against several others, then a `switch` statement can be more concise than the corresponding `if` statement. See `help switch` for more information.

6.5 The try/catch statement

Matrix computations can fail because of characteristics of the matrices that are hard to determine before doing the computation. If the failure is severe, your script or function (see Chapter 7) may be terminated. The `try/catch` statement allows you to compute optimistically and then recover if those computations fail. The general form is:

```
try
    statements
catch
    statements
end
```

The first block of statements is executed. If an error occurs, those statements are terminated, and the second block of statements is executed. You cannot do this with an `if` statement. See `doc try`. See Section 11.5 for an example of `try` and `catch`.

6.6 Matrix expressions (if and while)

A matrix expression is interpreted by `if` and `while` to be true if *every* entry of the matrix expression is nonzero. Enter these two matrices:

```
A = [ 1 2 ; 3 4 ]
B = [ 2 3 ; 3 5 ]
```

If you wish to execute a statement when matrices A and B are equal, you could type:

```
if A == B
    disp('A and B are equal')
end
```

If you wish to execute a statement when A and B are not equal, you would type:

```
if any(any(A ~= B))
    disp('A and B are not equal')
end
```

or, more simply,

```
if A == B else
    disp('A and B are not equal')
end
```

Note that the seemingly obvious:

```
if A ~= B
    disp('not what you think')
end
```

will not give what is intended because the statement would execute only if each of the corresponding entries of A and B differ. The functions any and all can be creatively used to reduce matrix expressions to vectors or scalars. Two anys are required above because any is a vector operator (see Section 5.3). In logical terms, any and all correspond to the existential ($\exists$) and universal ($\forall$) quantifiers, respectively, applied to each column of a matrix or each entry of a row or column vector. Like most vector functions, any and all can be applied to dimensions of a matrix other than the columns.

An `if` statement with a two-dimensional matrix *expression* is equivalent to:

```
if all(all(expression))
    statement
end
```

6.7 Infinite loops

With loops, it is possible to execute a command that will never stop. Typing Ctrl-C stops a runaway display or computation. Try:

```
i = 1
while i > 0
    i = i + 1
end
```

then type Ctrl-C to terminate this loop.

7. M-files

MATLAB can execute a sequence of statements stored in files. These are called M-files because they must have the file type `.m` as the last part of their filename.

7.1 M-file Editor/Debugger window

Much of your work with MATLAB will be in creating and refining M-files. M-files are usually created using your favorite text editor or with MATLAB's M-file Editor/Debugger. See also `Help: MATLAB: Desktop Tools and Development Environment: Editing and Debugging M-Files`.

There are two types of M-files: script files and function files. In this exercise, you will incrementally develop and debug a script and then a function for making a matrix diagonally dominant. Create a new M-file, either with the edit command, by selecting the File ▶ New ▶ M-file menu item, or by clicking the new-file button:

Type in these lines in the Editor,

```
f = sum(A, 2) ;
A = A + diag(f) ;
```

and save the file as **ddom.m** by clicking:

You have just created a MATLAB script file.[1] The semicolons are there because you normally do not want to see the results of every line of a script or function.

7.2 Script files

A script file consists of a sequence of normal MATLAB statements. Typing **ddom** in the Command window causes the statements in the script file **ddom.m** to be executed. Variables in a script file refer to variables in the main workspace, so changing them will change your workspace variables. Type:

```
A = rand(3)
ddom
A
```

---

[1] See http://www.cise.ufl.edu/research/sparse/MATLAB or http://www.crcpress.com for the M-files and MEX-files used in this book.

in the Command window. It seems to work; the matrix A is now diagonally dominant. If you type this in the Command window, though,

```
A = [1 -2 ; -1 1]
ddom
A
```

then the diagonal of A just got worse. What happened? Click on the Editor window and move the mouse to point to the variable f, anywhere in the script. You will see a yellow pop-up window with:

```
f =
    -1
     0
```

Oops. f is supposed to be a sum of absolute values, so it cannot be negative. Change the first line of ddom.m to:

```
f = sum(abs(A), 2) ;
```

save the file, and run it again on the original matrix A=[1 -2;-1 1]. This time, instead of typing in the command, try running the script by clicking:

in the Editor window. This is a shortcut to typing ddom in the Command window. The matrix A is now diagonally dominant. Run the script again, though, and you will see that A is modified even if it is already diagonally dominant. Fix this by modifying only those rows that violate diagonal dominance.

Set A to [1 -2;-1 1] by clicking on the command in the Command History window. Modify ddom.m to be:

```
d = diag(A) ;
a = abs(d) ;
f = sum(abs(A), 2) - a ;
i = find(f >= a) ;
A(i,i) = A(i,i) + diag(f(i)) ;
```

Save and run the script by clicking:

Run it again; the matrix does not change.

Try it on the matrix A=[-1 2;1 -1]. The result is wrong. To fix it, try another debugging method: setting breakpoints. A breakpoint causes the script to pause, and allows you to enter commands in the Command window, while the script is paused (it acts just like the keyboard command).

Click on line 5 and select Debug ▶ Set/Clear Breakpoint in the Editor or click:

A red dot appears in a column to the left of line 5. You can also set and clear breakpoints by clicking on the red dots or dashes in this column. In the Command window, type:

```
clear
A = [-1 2 ; 1 -1]
ddom
```

A green arrow appears at line 5, and the prompt K>>
appears in the Command window. Execution of the script
has paused, just before line 5 is executed. Look at the
variables A and f. Since the diagonal is negative, and f is
an absolute value, we should subtract f from A to
preserve the sign. Type the command:

```
A = A - diag(f)
```

The matrix is now correct, although this works only if all
of the rows need to be fixed and all diagonal entries are
negative. Stop the script by selecting Debug ▶ Exit
Debug Mode or by clicking:

Clear the breakpoint. Replace line 5 with:

```
s = sign(d(i)) ;
A(i,i) = A(i,i) + diag(s .* f(i)) ;
```

Type A=[-1 2;1 -1] and run the script. The script
seems to work, but it modifies A more than is needed. Try
the script on A=zeros(4), and you will see that the
matrix is not modified at all, because sign(0) is zero.
Fix the script so that it looks like this:

```
d = diag(A) ;
a = abs(d) ;
f = sum(abs(A), 2) - a ;
i = find(f >= a) ;
[m n] = size(A) ;
k = i + (i-1)*m ;
tol = 100 * eps ;
s = 2 * (d(i) >= 0) - 1 ;
A(k) = (1+tol) * s .* max(f(i), tol) ;
```

which is the code you typed in Section 5.6.

7.3 Function files

Function files provide extensibility to MATLAB. You can create new functions specific to your problem, which will then have the same status as other MATLAB functions. Variables in a function file are by default local. A variable can, however, be declared global (see doc global). Use global variables with caution; they can be a symptom of bad program design and can lead to hard-to-debug code.

Convert your ddom.m script into a function by adding these lines at the beginning of ddom.m:

```
function B = ddom(A)
% B = ddom(A) returns a diagonally
% dominant matrix B by modifying the
% diagonal of A.
```

and add this line at the end of your new function:

```
B = A ;
```

You now have a MATLAB function, with one input argument and one output argument. To see the difference between global and local variables as you do this exercise, type clear. Functions do not modify their inputs, so:

```
C = [1 -2 ; -1 1]
D = ddom(C)
```

returns a matrix D that is diagonally dominant. The matrix C in the workspace does not change, although a copy of it local to the ddom function, called A, is modified as the function executes. Note that the other variables, a, d, f, i, k and s no longer appear in your main workspace. Neither do A and B. These are local to the **ddom** function.

The first line of the function declares the function name, input arguments, and output arguments; without this line the file would be a script file. Then a MATLAB statement D=ddom(C), for example, causes the matrix C to be passed as the variable A in the function and causes the output result to be passed out to the variable D. Since variables in a function file are local, their names are independent of those in the current MATLAB workspace. Your workspace will have only the matrices C and D. If you want to modify C itself, then use C=ddom(C).

Lines that start with % are comments; more on this in Section 7.6. An optional **return** statement causes the function to finish and return its outputs. An M-file can reference other M-files, including itself recursively.

7.4 Multiple inputs and outputs

A function may also have multiple output arguments. For example, it would be useful to provide the caller of the **ddom** function some control over how strong the diagonal is to be and to provide more results, such as the list of rows (the variable i) that violated diagonal dominance. Try changing the first line to:

```
function [B,i] = ddom(A, tol)
```

and add a % at the beginning of the line that computes tol. Single assignments can also be made with a function having multiple output arguments. For example, with this version of **ddom**, the statement D=ddom(C,0.1)

41

will assign the modified matrix to the variable D without returning the vector i. Try it on C=[1 -2 ; -1 1].

7.5 Variable arguments

Not all inputs and outputs of a function need be present when the function is called. The variables nargin and nargout can be queried to determine the number of inputs and outputs present. For example, we could use a default tolerance if tol is not present. Add these statements in place of the line that computed tol:

```
if (nargin == 1)
    tol = 100 * eps ;
end
```

Section 8.1 gives an example of nargin and nargout.

7.6 Comments and documentation

The % symbol indicates that the rest of the line is a comment; MATLAB will ignore the rest of the line. Moreover, the first contiguous comment lines are used to document the M-file. They are available to the online help facility and will be displayed if help ddom or doc ddom are entered. Such documentation should always be included in a function file. Since you have modified the function to add new inputs and outputs, edit your script to describe the variables i and tol. Be sure to state what the default value of tol is. Next, type help ddom or doc ddom.

Block comments are useful for lengthy comments or for disabling code. A block comment starts with a line containing only %{ and ends with a line containing only %}. Block comments in an M-file are not printed by the help or doc commands.

A line starting with two percent signs (%%) denotes the beginning of a MATLAB code *cell*. This type of cell has nothing to do with cell arrays, but defines a section of code in an M-file. Cells can be executed by themselves, and cell publishing (discussed in Chapter 20) generates reports whose sections are defined by an M-file's cells.

7.7 MATLAB's path

M-files must be in a directory accessible to MATLAB. M-files in the current directory are always accessible. The current list of directories in MATLAB's search path is obtained by the command `path`. This command can also be used to add or delete directories from the search path. See `doc path`. The command `which` locates functions and files on the path. For example, type `which hilb`. You can modify your MATLAB path with the command `path`, or `pathtool`, which brings up another window. You can also select File ▸ Set Path.

8. Advanced M-file Features

This section describes advanced M-file techniques, such as how to pass a function as an argument to another function and how to write high-performance code in MATLAB.

8.1 Function handles and anonymous functions

A function handle (@) is a reference to a function that can then be treated as a variable. It can be copied, placed in cell array, and evaluated just like a regular function. For example,

```
f = @sqrt
f(2)
sqrt(2)
```

The `str2func` function converts a string to a function handle. For example,

```
f = str2func('sqrt')
f(2)
```

Function handles can refer to built-in MATLAB functions, to your own function in an M-file, or to anonymous functions. An anonymous function is defined with a one-line expression, rather than by an M-file. Try:

```
g = @(x) x^2-5*x+6-sin(9*x)
g(1)
```

Some MATLAB functions that operate on function handles need to evaluate the function on a vector, so it is often better to define an anonymous function (or M-file) so that it can operate entry-wise on scalars, vectors, or matrices. Try this instead:

```
g = @(x) x.^2-5*x+6-sin(9*x)
g(1)
```

The general syntax for an anonymous function is

handle = @(*arg1, arg2, ...*) *expression*

Here is an example with two input arguments:

```
norm2 = @(x,y) sqrt(x^2 + y^2)
norm2(4, 5)
norm([4 5])
```

One advantage of anonymous functions is that they can implicitly refer to variables in the workspace or the calling function without having to use the `global` statement. Try this example:

```
A = [3 2 ; 1 3]
b = [3 ; 4]
y = A\b
resid = @(x) A*x-b
resid(y)
A*y-b
```

In this case, x is an argument, but A and b are defined in the calling workspace.

To find out what a function handle refers to, use `func2str` or `functions`. Try these examples:

```
func2str(f)
func2str(g)
func2str(norm2)
func2str(resid)
functions(f)
```

Cell arrays can contain function handles. They can be indexed and the function evaluated in a single expression. Try this:

```
h{1} = f
h{2} = g
h{1}(2)
f(2)
h{2}(1)
g(1)
```

Here is a more useful example. The `bisect` function, below, solves the nonlinear equation $f(x)=0$. It takes a function handle or a string as one of its inputs. If the function is a string, it is converted to a function handle with `str2func`. `bisect` also gives you an example of `nargin` and `nargout` (see also Section 7.5). Compare `bisect` with the built-in `fzero` discussed in Section 18.4.

```
function [b, steps] = bisect(f,x,tol)
% BISECT:  zero of a function of one
% variable via the bisection method.
% bisect(f,x) returns a zero of the
% function f.  f is a function
% handle or a string with the name of a
% function.  x is an array of length 2;
% f(x(1)) and f(x(2)) must differ in
% sign.
%
% An optional third input argument sets
% a tolerance for the relative accuracy
% of the result.  The default is eps.
% An optional second output argument
% gives a matrix containing a trace of
% the steps; the rows are of the form
% [c f(c)].

if (nargin < 3)
    % default tolerance
    tol = eps ;
end
trace = (nargout == 2) ;
if (ischar(f))
    f = str2func(f) ;
end
a = x(1) ;
b = x(2) ;
fa = f(a) ;
fb = f(b) ;
if (trace)
    steps = [a fa ; b fb] ;
end
% main loop
while (abs(b-a) > 2*tol*max(abs(b),1))
    c = a + (b-a)/2 ;
    fc = f(c) ;
    if (trace)
        steps = [steps ; [c fc]] ;
    end
```

46

```
      if (fb > 0) == (fc > 0)
            b = c ;
            fb = fc ;
      else
            a = c ;
            fa = fc ;
      end
end
```

Type in `bisect.m`, and then try:

```
bisect(@sin, [3 4])
bisect('sin', [3 4])
bisect(g, [0 3])
g(ans)
```

Some of MATLAB's functions are built in; others are distributed as M-files. The actual listing of any M-file, MATLAB's or your own, can be viewed with the MATLAB command `type`. Try entering `type eig`, `type vander`, and `type rank`.

8.2 Name resolution

When MATLAB comes upon a new name, it resolves it into a specific variable or function by checking to see if it is a variable, a built-in function, a file in the current directory, or a file in the MATLAB path (in order of the directories listed in the path). MATLAB uses the first variable, function, or file it encounters with the specified name. There are other cases; see Help: MATLAB: Desktop Tools and Development Environment: Workspace, Search Path, and File Operations: Search Path. You can use the command `which` to find out what a name is. Try this:

```
clear
i
which i
```

```
i = 3
which i
```

8.3 Error and warning messages

Error messages are best displayed with the function
`error`. For example,

```
A = rand(4,3)
[m n] = size(A) ;
if m ~= n
    error('A must be square') ;
end
```

aborts execution of an M–file if the matrix A is not
square. This is a useful thing to add to the `ddom` function
that you developed in Chapter 7, since diagonal
dominance is only defined for square matrices. Try
adding it to `ddom` (excluding the `rand` statement, of
course), and see what happens if you call `ddom` with a
rectangular matrix.

If you want to print a warning, but continue execution,
use the `warning` statement instead, as in:

```
warning('A singular; computing anyway')
```

The warning function can also turn on or off the warnings
that MATLAB provides. If you know that a divide by
zero is safe in your application, use

```
warning('off', 'MATLAB:divideByZero')
```

Try computing 1/0 both before and after you type in the
above `warning` statement. Use `'on'` in the first
argument to turn the warning back on for subsequent

division by zero. `warning`, with no arguments, displays
a list of disabled warnings.

See Section 6.5 (`try/catch`) for one way to deal with
errors in functions you call.

8.4 User input

In an M-file the user can be prompted to interactively
enter input data, expressions, or commands. When, for
example, the statement:

```
iter = input('iteration count: ') ;
```

is encountered, the prompt message is displayed and
execution pauses while the user keys in the input data (or,
in general, any MATLAB expression). Upon pressing the
return or entry key, the data is assigned to the variable
`iter` and execution resumes. You can also input a string;
see `help input`.

An M-file can be paused until a return is typed in the
Command window with the `pause` command. It is a
good idea to display a message, as in:

```
disp('Hit enter to continue: ') ;
pause
```

A Ctrl-C will terminate the script or function that is
paused. A more general command, `keyboard`, allows
you to type any number of MATLAB commands. See
`doc keyboard`.

8.5 Performance measures

Time and space are the two basic measures of an
algorithm's efficiency. In MATLAB, this translates into

the number of floating-point operations (flops) performed, the elapsed time, the CPU time, and the memory space used. MATLAB no longer provides a flop count because it uses high-performance block matrix algorithms that make it difficult to count the actual flops performed. On current computers with deep memory hierarchies, flop count is less useful as a performance predictor than it once was. See help flops.

The elapsed time (in seconds) can be obtained with tic and toc; tic starts the timer and toc returns the elapsed time since the last tic. Hence:

```
tic ; statement ; t = toc
```

will return the elapsed time t for execution of the *statement*. Type it as one line in the Command window. Otherwise, the timer records the time you took to type the statement. The elapsed time for solving a linear system above can be obtained, for example, with:

```
n = 1000 ;
A = rand(n) ;
b = rand(n,1) ;
tic ; x = A\b ; t = toc
r = norm(A*x-b)
(2/3)*n^3 / t
```

The norm of the residual is also computed, and the last line reports the approximate flop rate. You may wish to compare x=A\b with x=inv(A)*b for solving the linear system. Try it. You will generally find A\b to be faster and more accurate.

If there are other programs running at the same time on your computer, elapsed time will not be an accurate measure of performance. Try using `cputime` instead. See `doc cputime`.

MATLAB runs faster if you can restructure your computations to use less memory. Type the following and select n to be some large integer, such as:

```
n = 16000 ;
a = rand(n,1) ;
b = rand(1,n) ;
c = rand(n,1) ;
```

Here are three ways of computing the same vector x. The first one uses hardly any extra memory, the second and third use a huge amount. Try them:

```
x = a*(b*c) ;
x = (a*b)*c ;
x = a*b*c ;
```

No measure of peak memory usage is provided. You can find out the total size of your workspace, in bytes, with the command `whos`. The total can also be computed:

```
s = whos
space = sum([s.bytes])
```

Try it. This does not give the peak memory used while inside a MATLAB operator or function, though. Type `doc memory` for more memory usage options.

8.6 Efficient code

The function `ddom.m` that you wrote in Chapter 7 and 8 illustrates some of the MATLAB features that can be used to produce efficient code. All operations are

"vectorized," and loops are avoided. We could have
written the ddom function using nested `for` loops:

```
function B = ddomloops(A,tol)
% B = ddomloops(A) returns a
% diagonally dominant matrix B by modifying
% the diagonal of A.
[m n] = size(A) ;
if (nargin == 1)
    tol = 100 * eps ;
end
for i = 1:n
    d = A(i,i) ;
    a = abs(d) ;
    f = 0 ;
    for j = 1:n
        if (i ~= j)
            f = f + abs(A(i,j)) ;
        end
    end
    if (f >= a)
        aii = (1 + tol) * max(f, tol) ;
        if (d < 0)
            aii = -aii ;
        end
        A(i,i) = aii ;
    end
end
B = A ;
```

The non-vectorized **ddomloops** function is only slightly
slower than the vectorized **ddom**. In earlier versions of
MATLAB, the non-vectorized version would be very
slow. MATLAB 6.5 and subsequent versions include an
accelerator that greatly improves the performance of non-
vectorized code. Try:

```
A = rand(1000) ;
tic ; B = ddom(A) ; toc
tic ; B = ddomloops(A) ; toc
```

Only simple `for` loops can be accelerated. Loops that
operate on sparse matrices are not accelerated, for

example (sparse matrices are discussed in Chapter 15). Try:

```
A = sparse(A) ;
tic ; B = ddom(A) ; toc
tic ; B = ddomloops(A) ; toc
```

Since not every loop can be accelerated, writing code that has no for or while loops is still important. As you become practiced in writing without loops and reading loop-free MATLAB code, you will also find that the loop-free version is often easier to read and understand.

If you cannot vectorize a loop, you can speed it up by preallocating any vectors or matrices in which output is stored. For example, by including the second statement below, which uses the function zeros, space for storing E in memory is preallocated. Without this, MATLAB must resize E one column larger in each iteration, slowing execution.

```
M = magic(6) ;
E = zeros(6,50) ;
for j = 1:50
    E(:,j) = eig(M^j) ;
end
```

9. Calling C from MATLAB

There are times when MATLAB itself is not enough. You may have a large application or library written in another language that you would like to use from MATLAB, or it might be that the performance of your M-file is not what you would like.

MATLAB can call routines written in C, Fortran, or Java. Similarly, programs written in C and Fortran can call

MATLAB. In this chapter, we will just look at how to call a C routine from MATLAB. For more information, see Help: MATLAB: External Interfaces, or see the online MATLAB documents *External Interfaces* and *External Interfaces Reference*. This discussion assumes that you already know C.

9.1 A simple example

A routine written in C that can be called from MATLAB is called a MEX-file. The routine must always have the name mexFunction, and the arguments to this routine are always the same. Here is a very simple MEX-file; type it in as the file hello.c in your favorite text editor.

```c
#include "mex.h"
void mexFunction
(
    int nargout,
    mxArray *pargout [ ],
    int nargin,
    const mxArray *pargin [ ]
)
{
    mexPrintf ("hello world\n") ;
}
```

Compile and run it by typing:

```
mex hello.c
hello
```

If this is the first time you have compiled a C MEX-file on a PC with Microsoft Windows, you will be prompted to select a C compiler. MATLAB for the PC comes with its own C compiler (lcc). The arguments nargout and nargin are the number of outputs and inputs to the function (just as an M-file function), and pargout and pargin are pointers to the arguments themselves (of type mxArray). This `hello.c` MEX-file does not have any inputs or outputs, though.

The `mexPrintf` function is just the same as `printf`. You can also use `printf` itself; the `mex` command redefines it as `mexPrintf` when the program is compiled with a `#define`. This way, you can write a routine that can be used from MATLAB or from a stand-alone C application, without MATLAB.

9.2 C versus MATLAB arrays

MATLAB stores its arrays in column major order, while the convention for C is to store them in row major order. Also, the number of columns in an array is not known until the mexFunction is called. Thus, two-dimensional arrays in MATLAB must be accessed with one-dimensional indexing in C (see also Section 5.6). In the example in the next section, the `INDEX` macro helps with this translation.

Array indices also appear differently. MATLAB is written in C, and it stores all of its arrays internally using zero-based indexing. An m-by-n matrix has rows 0 to m$-$1 and columns 0 to n$-$1. However, the user interface to these arrays is always one-based, and index vectors in MATLAB are always one-based. In the example below, one is added to the `List` array returned by `diagdom` to account for this difference.

9.3 A matrix computation in C

In Chapters 7 and 8, you wrote the function `ddom.m`. Here is the same function written as an ANSI C MEX-file. Compare the `diagdom` routine with the loop-based version ddomloops.m in Section 8.6. MATLAB mx and mex routines are described in Section 9.4.

```c
#include "mex.h"
#include "matrix.h"
#include <stdlib.h>
#include <float.h>
#define INDEX(i,j,m) ((i)+(j)*(m))
#define ABS(x) ((x) >= 0 ? (x) : -(x))
#define MAX(x,y) (((x)>(y)) ? (x):(y))

void diagdom
(
    double *A, int n, double *B,
    double tol, int *List, int *nList
)
{
    double d, a, f, bij, bii ;
    int i, j, k ;
    for (k = 0 ; k < n*n ; k++)
    {
        B [k] = A [k] ;
    }
    if (tol < 0)
    {
        tol = 100 * DBL_EPSILON ;
    }
    k = 0 ;
    for (i = 0 ; i < n ; i++)
    {
        d = B [INDEX (i,i,n)] ;
        a = ABS (d) ;
        f = 0 ;
        for (j = 0 ; j < n ; j++)
        {
            if (i != j)
            {
                bij = B [INDEX (i,j,n)] ;
                f += ABS (bij) ;
            }
        }
        if (f >= a)
        {
            List [k++] = i ;
            bii = (1 + tol) * MAX (f, tol) ;
            if (d < 0)
            {
```

```
                bii = -bii ;
            }
            B [INDEX (i,i,n)] = bii ;
        }
    }
    *nList = k ;
}

void error (char *s)
{
    mexPrintf
    ("Usage: [B,i] = diagdom (A,tol)\n") ;
    mexErrMsgTxt (s) ;
}

void mexFunction
(
    int nargout, mxArray *pargout [ ],
    int nargin,  const mxArray *pargin [ ]
)
{
    double tol, *A, *B, *I ;
    int n, k, *List, nList ;

    /* get inputs A and tol */
    if (nargout > 2 || nargin > 2 || nargin==0)
    {
        error ("Wrong number of arguments") ;
    }
    if (mxIsSparse (pargin [0]))
    {
        error ("A cannot be sparse") ;
    }
    n = mxGetN (pargin [0]) ;
    if (n != mxGetM (pargin [0]))
    {
        error ("A must be square") ;
    }
    A = mxGetPr (pargin [0]) ;
    tol = -1 ;
    if (nargin > 1)
    {
        if (!mxIsEmpty (pargin [1]) &&
            mxIsDouble (pargin [1]) &&
            !mxIsComplex (pargin [1]) &&
            mxIsScalar (pargin [1]))
        {
            tol = mxGetScalar (pargin [1]) ;
        }
```

57

```
            else
            {
                error ("tol must be scalar") ;
            }
    }

    /* create output B */
    pargout [0] =
        mxCreateDoubleMatrix (n, n, mxREAL) ;
    B = mxGetPr (pargout [0]) ;

    /* get temporary workspace */
    List = (int *) mxMalloc (n * sizeof (int)) ;

    /* do the computation */
    diagdom (A, n, B, tol, List, &nList) ;

    /* create output I */
    pargout [1] =
        mxCreateDoubleMatrix (nList, 1, mxREAL);
    I = mxGetPr (pargout [1]) ;
    for (k = 0 ; k < nList ; k++)
    {
        I [k] = (double) (List[k] + 1) ;
    }

    /* free the workspace */
    mxFree (List) ;
}
```

Type it in as the file diagdom.c (or get it from the web),
and then type:

```
mex diagdom.c
A = rand(6) ;
B = ddom(A) ;
C = diagdom(A) ;
```

The matrices B and C will be the same (round-off error
might cause them to differ slightly). The C mexFunction
diagdom is about 3 times faster than the M-file ddom for
large matrices.

9.4 MATLAB mx and mex routines

In the last example, the C routine calls several MATLAB routines with the prefix mx or mex. Routines with mx prefixes operate on MATLAB matrices and include:

mxIsEmpty	1 if the matrix is empty, 0 otherwise
mxIsSparse	1 if the matrix is sparse, 0 otherwise
mxGetN	number of columns of a matrix
mxGetM	number of rows of a matrix
mxGetPr	pointer to the real values of a matrix
mxGetScalar	the value of a scalar
mxCreateDoubleMatrix	create MATLAB matrix
mxMalloc	like malloc in ANSI C
mxFree	like free in ANSI C

Routines with mex prefixes operate on the MATLAB environment and include:

mexPrintf	like printf in C
mexErrMsgTxt	like MATLAB's error statement
mexFunction	the gateway routine from MATLAB

You will note that all of the references to MATLAB's mx and mex routines are limited to the mexFunction gateway routine. This is not required; it is just a good idea. Many other mx and mex routines are available.

The memory management routines in MATLAB (mxMalloc, mxFree, and mxCalloc) are much easier to use than their ANSI C counterparts. If a memory allocation request fails, the mexFunction terminates and control is passed backed to MATLAB. Any workspace allocated by mxMalloc that is not freed when the mexFunction returns or terminates is automatically freed by MATLAB. This is why no memory allocation error checking is included in `diagdom.c`; it is not necessary.

9.5 Online help for MEX routines

Create an M-file called `diagdom.m`, with only this:

```
function [B,i] = diagdom(A,tol)
%DIAGDOM:  modify the matrix A
% [B,i] = diagdom(A,tol) returns a
% diagonally dominant matrix B by
% modifying the diagonal of A.  i is a
% list of modified diagonal entries.
error('diagdom mexFunction not found');
```

Now type `help diagdom` or `doc diagdom`. This is a simple method for providing online help for your own MEX-files.

If both `diagdom.m` and the compiled `diagdom` mexFunction are in MATLAB's path, then the `diagdom` mexFunction is called. If only the M-file is in the path, it is called instead; thus the error statement in `diagdom.m` above.

9.6 Larger examples on the web

The `colamd` and `symamd` routines in MATLAB are C MEX-files. The source code for these routines is on the web at http://www.cise.ufl.edu/research/sparse/colamd. Like the example in the previous section, they are split into a `mexFunction` gateway routine and another set of routines that do not make use of MATLAB. A simpler example is a sparse LDL^T factorization routine that takes less memory than MATLAB's `chol`, at http://www.cise.ufl.edu/research/sparse/ldl.

10. Calling Fortran from MATLAB

C is a great language for numerical calculations, particularly if the data structures are complicated. MATLAB itself is written in C. No single language is best for all tasks, however, and that holds for C as well. In this chapter we will look at how to call a Fortran subroutine from MATLAB. A Fortran subroutine is accessed via a mexFunction in much the same way as a C subroutine is called. Normally, the mexFunction acts as a gateway routine that gets its input arguments from MATLAB, calls a computational routine, and then returns its output arguments to MATLAB, just like the C example in the previous chapter.

10.1 Solving a transposed system

The `linsolve` function was introduced in Section 5.5. Here is a Fortran subroutine `utsolve` that computes x=A'\b when A is dense, square, real, and upper triangular. It has no calls to MATLAB-specific mx or mex routines.

```
      subroutine utsolve (n, x, A, b)
      integer n
      real*8 x(n), A(n,n), b(n), xi
      integer i, j
      do 1 i = 1,n
        xi = b(i)
        do 2 j = 1,i-1
          xi = xi - A(j,i) * x(j)
 2      continue
        x(i) = xi / A(i,i)
 1    continue
      return
      end
```

10.2 A Fortran mexFunction with %val

To call this computational subroutine from MATLAB as
x=utsolve(A,b), we need a gateway routine, the first
lines of which must be:

```
      subroutine mexFunction
     $ (nargout, pargout, nargin, pargin)
      integer nargout, nargin
      integer pargout (*), pargin (*)
```

where the $ is in column 6. These lines must be the same
for any Fortran mexFunction (you do not need to split
the first line). Note that pargin and pargout are arrays
of integers. MATLAB passes its inputs and outputs as
pointers to objects of type mxArray, but Fortran cannot
handle pointers. Most Fortran compilers can convert
integer "pointers" to references to Fortran arrays via the
non-standard %val construct. We will use this in our
gateway routine. The next two lines of the gateway
routine declare some MATLAB mx-routines.

```
      integer mxGetN, mxGetPr
      integer mxCreateDoubleMatrix
```

This is required because Fortran has no include-file
mechanism. The next lines determine the size of the
input matrix and create the n-by-1 output vector x.

```
      integer n
      n = mxGetN (pargin (1))
      pargout (1) =
     $ mxCreateDoubleMatrix (n, 1, 0)
```

We can now convert "pointers" into Fortran array
references and call the computational routine.

```
      call utsolve (n,
$ %val (mxGetPr (pargout (1))),
$ %val (mxGetPr (pargin (1))),
$ %val (mxGetPr (pargin (2))))
      return
      end
```

The arrays in both MATLAB and Fortran are column-
oriented and one-based, so translation is not necessary as
it was in the C mexFunction.

Combine the two routines into a single file called
utsolve.f and type:

```
    mex utsolve.f
```

in the MATLAB command window. Error checking
could be added to ensure that the two input arguments are
of the correct size and type. The code would look much
like the C example in Chapter 9, so it is not included.
Test this routine on as large a matrix that your computer
can handle.

```
    n = 5000
    A = triu(rand(n,n)) ;
    b = rand(n,1) ;
    tic ; x = A'\b ; toc
    opts.UT = true
    opts.TRANSA = true
    tic ; x2 = linsolve(A,b,opts) ; toc
    tic ; x3 = utsolve(A,b) ; toc
    norm(x-x2)
    norm(x-x3)
```

The solutions should agree quite closely. On a Pentium 4,
both linsolve and utsolve are about 15 times faster
than x=A'\b. They require less memory, as well, since
they do not have to construct A'.

10.3 If you cannot use %val

If your Fortran compiler does not support the %val construct, then you will need to call MATLAB mx-routines to copy the MATLAB arrays into Fortran arrays, and vice versa. The GNU f77 compiler supports %val, but issues a warning that you can safely ignore.

In this utsolve example, add this to your mexFunction gateway routine:

```
integer nmax
parameter (nmax = 5000)
real*8 A(nmax,nmax), x(nmax), b(nmax)
```

where nmax is the largest dimension you want your function to handle. Unless you want to live dangerously, you should check n to make sure it is not too big:

```
if (n .gt. nmax) then
    call mexErrMsgTxt ("n too big")
endif
```

Replace the call to utsolve with this code.

```
call mxCopyPtrToReal8
$ (mxGetPr (pargin (1)), A, n**2)
call mxCopyPtrToReal8
$ (mxGetPr (pargin (2)), b, n)
call lsolve (n, x, A, b)
call mxCopyReal8ToPtr
$ (x, mxGetPr (pargout (1)), n)
```

This copies the input MATLAB arrays A and b to their Fortran counterparts, calls the utsolve routine, and then copies the solution x to its MATLAB counterpart. Although this is more portable, it takes more memory and is significantly slower. If possible, use %val.

11. Calling Java from MATLAB

While C and Fortran excel at numerical computations,
Java is well-suited to web-related applications and
graphical user interfaces. MATLAB can handle native
Java objects in its workspace and can directly call Java
methods on those objects. No mexFunction is required.

11.1 A simple example

Try this in the MATLAB Command window

```
t = 'hello world'
s = java.lang.String(t)
s.indexOf('w') + 1
find(s == 'w')
whos
```

You have just created a Java string in the MATLAB
workspace, and determined that the character 'w' appears
as the seventh entry in the string using both the indexOf
Java method and the find MATLAB function.

11.2 Encryption/decryption

MATLAB can handle strings on its own, without help
from Java, of course. Here is a more interesting example.
Type in the following as the M-file getkey.m.

```
function key = getkey(password)
%GETKEY: key = getkey(password)
% Converts a string into a key for use
% in the encrypt and decrypt functions.
% Uses triple DES.
import javax.crypto.spec.*
b = int8(password) ;
n = length(b) ;
b((n+1):24) = 0 ;
b = b(1:24) ;
key = SecretKeySpec(b, 'DESede') ;
```

The getkey routine takes a password string and converts it into a 24-byte triple DES key using the javax.crypto package. You can then encrypt a string with the encrypt function:

```
function e = encrypt(t, key)
%ENCRYPT: e = encrypt(t, key)
% Encrypt the plaintext string t into
% the encrypted byte array e using a key
% from getkey.
import javax.crypto.*
cipher = Cipher.getInstance('DESede') ;
cipher.init(Cipher.ENCRYPT_MODE, key) ;
e = cipher.doFinal(int8(t))' ;
```

Except for the function statement and the comments, this looks almost exactly the same as the equivalent Java code. This is not a Java program, however, but a MATLAB M-file that uses Java objects and methods. Finally, the decrypt function undoes the encryption: .

```
function t = decrypt(e, key)
%DECRYPT: t = decrypt(e, key)
% Decrypt the encrypted byte array e
% into to plaintext string t using a key
% from getkey.
import javax.crypto.*
cipher = Cipher.getInstance('DESede') ;
cipher.init(Cipher.DECRYPT_MODE, key) ;
t = char(cipher.doFinal(e))' ;
```

With these three functions in place, try:

```
k = getkey('this is a secret password')
e = encrypt('a hidden message',k)
decrypt(e,k)
```

Now you encrypt and decrypt strings in MATLAB.

66

11.3 MATLAB's Java class path

If you define your own Java classes that you want to use within MATLAB, you need to modify your Java class path. This path is different than the path used to find M-files. You can add directories to the static Java path by editing the file `classpath.txt`, or you can add them to your dynamic Java path with the command

    javaaddpath *directory*

where *directory* is the name of a directory containing compiled Java classes. `javaclasspath` lists the directories where MATLAB looks for Java classes.

If you do not have write permission to `classpath.txt`, you need to type the `javaaddpath` command every time you start MATLAB. You can do this automatically by creating an M-file script called `startup.m` and placing in it the `javaaddpath` command. Place this file in one of the directories in your MATLAB path, or in the directory in which you launch MATLAB, and it will be executed whenever MATLAB starts.

11.4 Calling your own Java methods

To write your own Java classes that you can call from MATLAB, you must first download and install the Java 2 SDK (Software Development Kit) Version 1.4 (or later) from java.sun.com. Edit your operating system's PATH variable so that you can type the command `javac` in your operating system command prompt.

MATLAB includes two M-files that can download a web page into either a string (`urlread`) or a file (`urlwrite`). Try:

```
s = urlread('http://www.mathworks.com')
```

The urlread function is an M-file. You can take a look
at it with the command edit urlread. It uses a Java
package from The MathWorks called
mlwidgets.io.InterruptibleStreamCopier, but
only the compiled class file is distributed, not the Java
source file. Create your own URL reader, a purely Java
program, and put it in a file called myreader.java:

```java
import java.io.* ;
import java.net.* ;
public class myreader
{
  public static void main (String [ ] args)
  {
    geturl (args [0], args [1]) ;
  }
  public static void geturl (String u, String f)
  {
    try
    {
      URL url = new URL (u) ;
      InputStream i = url.openStream ();
      OutputStream o = new FileOutputStream (f);
      byte [ ] s = new byte [4096] ;
      int b ;
      while ((b = i.read (s)) != -1)
      {
        o.write (s, 0, b) ;
      }
      i.close ( ) ;
      o.close ( ) ;
    }
    catch (Exception e)
    {
      System.out.println (e) ;
    }
  }
}
```

The geturl method opens the URL given by the string
u, and copies it into a file whose name is given by the

string f. In either Linux/Unix or Windows, you can
compile this Java program and run it by typing these
commands at your operating system command prompt:

```
javac myreader.java
java myreader http://www.google.com my.txt
```

The second command copies Google's home page into
your own file called my.txt. You can also type the
commands in the MATLAB Command window, as in:

```
!javac myreader.java
```

Now that you have your own Java method, you can call it
from MATLAB just as the java.lang.String and
javax.crypto.* methods. In the MATLAB command
window, type (as one line):

```
myreader.geturl
('http://www.google.com','my.txt')
```

11.5 Loading a URL as a matrix

An even more interesting use of the myreader.geturl
method is to load a MAT-file or ASCII file from a web
page directly into the MATLAB workspace as a matrix.
Here is a simple loadurl M-file that does just that. It
can read compressed files; the Java method uncompresses
the URL automatically if it is compressed.

```
function result = loadurl(url)
% result = loadurl(url)
% Reads the URL given by the input
% string, url, into a temporary file
% using myread.java, loads it into a
% MATLAB variable, and returns the
% result. The URL can contain a MATLAB-
% readable text file, or a MAT-file.
t = tempname ;
```

```
myreader.geturl(url, t) ;
% load the temporary file, if it exists
try
    % try loading as an ascii file first
    result = load(t) ;
catch
    % try as a MAT file if ascii fails
    try
        result = load('-mat', t) ;
    catch
        result = [ ] ;
    end
end
% delete the temporary file
if (exist(t, 'file'))
    delete(t) ;
end
```

Try it with a simple text file (type this in as one line):

```
w = loadurl('http://www.cise.ufl.edu/
research/sparse/MATLAB/w')
```

which loads in a 2-by-2 matrix. Also try it with this
rather lengthy URL (type the string on one line):

```
s = loadurl('http://www.cise.ufl.edu/
research/sparse/mat/HB/west0479.mat.gz')
prob = s.Problem
spy(prob.A)
title([prob.name ': ' prob.title])
```

MATLAB 7.0 can create compressed MAT-files, so in
the future you may need to exclude the .gz extension in
this URL. spy plots a sparse matrix (see Section 15.5).

12. Two-Dimensional Graphics

MATLAB can produce two-dimensional plots. The
primary command for this is plot. Chapter 13 discusses
three-dimensional graphics. To preview some of these
capabilities, enter the command demo and select some of

the visualization and graphics demos. See Chapter 16 for a discussion of how to plot symbolic functions. Just like any other window, a Figure window can be docked in the main MATLAB window (except on the Macintosh).

12.1 Planar plots

The plot command creates linear x–y plots; if x and y are vectors of the same length, the command plot(x,y) opens a graphics window and draws an x–y plot of the elements of y versus the elements of x. You can, for example, draw the graph of the sine function over the interval -4 to 4 with the following commands:

```
x = -4:0.01:4 ;
y = sin(x) ;
plot(x, y) ;
```

Try it. The vector x is a partition of the domain with mesh size 0.01, and y is a vector giving the values of sine at the nodes of this partition (recall that sin operates entry-wise). When plotting a curve, the plot routine is actually connecting consecutive points induced by the partition with line segments. Thus, the mesh size should be chosen sufficiently small to render the appearance of a smooth curve.

The next example draws the graph of $y = e^{-x^2}$ over the interval -3 to 3. Note that you must precede ∧ by a period to ensure that it operates entry-wise:

```
x = -3:.01:3 ;
y = exp(-x.^2) ;
plot(x, y) ;
```

Select Tools ▶ Zoom In or Tools ▶ Zoom Out in the
Figure window to zoom in or out, or click these buttons
(or see the zoom command):

🔍 🔍

12.2 Multiple figures

You can have several concurrent Figure windows, one of
which will at any time be the designated current figure in
which graphs from subsequent plotting commands will be
placed. If, for example, Figure 1 is the current figure,
then the command figure(2) (or simply figure) will
open a second figure (if necessary) and make it the
current figure. The command figure(1) will then
expose Figure 1 and make it again the current figure. The
command gcf returns the current figure number, and
figure(gcf) brings the current figure window up.

MATLAB does not draw a plot right away. It waits until
all computations are finished, until a figure command is
encountered, or until the script or function requests user
input (see Section 8.4). To force MATLAB to draw a
plot right away, use the command drawnow. This does
not change the current figure.

12.3 Graph of a function

MATLAB supplies a function fplot to plot the graph of
a function. For example, to plot the graph of the function
above, you can first define the function in an M-file
called, say, expnormal.m containing:

```
function y = expnormal(x)
y = exp(-x.^2) ;
```

Then:

```
fplot(@expnormal, [-3 3])
```

will produce the graph over the indicated *x*-domain.

Using an anonymous function gives the same result without creating expnormal.m:

```
f = @(x) exp(-x.^2)
fplot(f, [-3 3])
```

12.4 Parametrically defined curves

Plots of parametrically defined curves can also be made:

```
t = 0:.001:2*pi ;
x = cos(3*t) ;
y = sin(2*t) ;
plot(x, y) ;
```

12.5 Titles, labels, text in a graph

The graphs can be given titles, axes labeled, and text placed within the graph with the following commands, which take a string as an argument.

title	graph title
xlabel	*x*-axis label
ylabel	*y*-axis label
gtext	place text on graph using the mouse
text	position text at specified coordinates

For example, the command:

```
title('A parametric cos/sin curve')
```

gives a graph a title. The command gtext('The Spot') lets you interactively place the designated text on the current graph by placing the mouse crosshair at the desired position and clicking the mouse. It is a good idea to prompt the user before using gtext. To place text in a graph at designated coordinates, use the command text (see doc text). These commands are also in the Insert menu in the Figure window. Select Insert ▸ TextBox, click on the figure, type something, and then click somewhere else to finish entering the text. If the edit-figure button ⌖ is depressed (or select Tools ▸ Edit Plot), you can right-click on anything in the figure and see a pop-up menu that gives you options to modify the item you just clicked. You can click and drag objects on the figure. Selecting Edit ▸ Axes Properties brings up a window with many more options. For example, clicking the Grid: ☑ X ☑ Y boxes adds grid lines (as does the grid command).

12.6 Control of axes and scaling

By default, MATLAB scales the axes itself (auto-scaling). This can be overridden by the command axis or by selecting Edit ▸ Axes Properties. Some features of the axis command are:

```
axis([xmin xmax ymin ymax])
                     sets the axes
axis manual          freezes the current axes for
                     new plots
axis auto            returns to auto-scaling
v = axis             vector v shows current scaling
axis square          axes same size (but not scale)
axis equal           same scale and tic marks on axes
```

| axis off | removes the axes |
| axis on | restores the axes |

The axis command should be given after the plot command. Try axis([-2 2 -3 3]) with the current figure. You will note that text entered on the figure using the text or gtext moves as the scaling changes (think of it as attached to the data you plotted). Text entered via Insert ▸ TextBox stays put.

12.7 Multiple plots

Here is one way to make multiple plots on a single graph:

```
x = 0:.01:2*pi;
y1 = sin(x) ;
y2 = sin(2*x) ;
y3 = sin(4*x) ;
plot(x, y1, x, y2, x, y3)
```

Another method uses a matrix Y containing the functional values as columns:

```
x = (0:.01:2*pi)' ;
y = [sin(x), sin(2*x), sin(4*x)] ;
plot(x, y)
```

The x and y pairs must have the same length, but each pair can have different lengths. Try:

```
plot(x, y, [0 2*pi], [0 0])
```

The command hold on freezes the current graphics screen so that subsequent plots are superimposed on it. The axes may, however, become rescaled. Entering hold off releases the hold. clf clears the figure. legend places a legend in the current figure to identify the different graphs. See doc legend.

12.8 Line types, marker types, colors

You can override the default line types, marker types, and colors. For example,

```
x = 0:.01:2*pi ;
y1 = sin(x) ;
y2 = sin(2*x) ;
y3 = sin(4*x) ;
plot(x,y1, '--', x,y2, ':', x,y3, 'o')
```

renders a dashed line and dotted line for the first two graphs, whereas for the third the symbol o is placed at each node. The line types are:

'-'	solid	':'	dotted
'--'	dashed	'-.'	dashdot

and the marker types are:

'.'	point	'o'	circle
'x'	x-mark	'+'	plus
'*'	star	's'	square
'd'	diamond	'v'	triangle-down
'^'	triangle-up	'<'	triangle-left
'>'	triangle-right	'p'	pentagram
'h'	hexagram		

Colors can be specified for the line and marker types:

'y'	yellow	'm'	magenta
'c'	cyan	'r'	red
'g'	green	'b'	blue
'w'	white	'k'	black

Thus, `plot(x,y1,'r--')` plots a red dashed line.

12.9 Subplots and specialized plots

The command `subplot(m,n,p)` partitions a single
figure into an m-by-n array of panes, and makes pane p
the current plot. The panes are numbered left to right. A
subplot can span multiple panes by specifying a vector p.
Here the last example, with each data set plotted in a
separate subplot:

```
subplot(2,2,1)
plot(x,y1, '--')
subplot(2,2,2)
plot(x,y2, ':')
subplot(2,2,[3 4])
plot(x,y3, 'o')
```

Other specialized planar plotting functions you may wish
to explore via `help` are:

```
bar       fill     quiver
compass   hist     rose
feather   polar    stairs
```

12.10 Graphics hard copy

Select `File ▶ Print` or click the print button

in the Figure window to send a copy of your figure to
your default printer. Layout options and selecting a
printer can be done with `File ▶ Page Setup` and `File ▶`
`Print Setup`.

You can save the figure as a file for later use in a
MATLAB Figure window. Try the save button

or File ▶ Save. This saves the figure as a .fig file, which can be later opened in the Figure window with the open button

or with File ▶ Open. Selecting File ▶ Export Setup or File ▶ Save As allows you to convert your figure to many other formats.

13. Three-Dimensional Graphics

MATLAB's primary commands for creating three-dimensional graphics of numerically-defined functions are plot3, mesh, surf, and light. Plotting of symbolic functions is discussed in Chapter 16. The menu options and commands for setting axes, scaling, and placing text, labels, and legends on a graph also apply for 3-D graphs. A zlabel can be added. The axis command requires a vector of length 6 with a 3-D graph.

13.1 Curve plots

Completely analogous to plot in two dimensions, the command plot3 produces curves in three-dimensional space. If x, y, and z are three vectors of the same size, then the command plot3(x,y,z) produces a perspective plot of the piecewise linear curve in three-space passing through the points whose coordinates are the respective elements of x, y, and z. These vectors are usually defined parametrically. For example,

```
t = .01:.01:20*pi ;
x = cos(t) ;
```

```
y = sin(t) ;
z = t.^3 ;
plot3(x, y, z)
```

produces a helix that is compressed near the *x-y* plane (a "slinky"). Try it.

13.2 Mesh and surface plots

The mesh command draws three-dimensional wire mesh surface plots. The command mesh(z) creates a three-dimensional perspective plot of the elements of the matrix z. The mesh surface is defined by the *z*-coordinates of points above a rectangular grid in the *x-y* plane. Try mesh(eye(20)).

Similarly, three-dimensional faceted surface plots are drawn with the command surf. Try surf(eye(20)).

To draw the graph of a function $z = f(x, y)$ over a rectangle, first define vectors xx and yy, which give partitions of the sides of the rectangle. The function [x,y]=meshgrid(xx,yy) then creates a matrix x, each row of which equals xx (whose column length is the length of yy) and similarly a matrix y, each column of which equals yy. A matrix z, to which mesh or surf can be applied, is then computed by evaluating the function f entry-wise over the matrices x and y.

You can, for example, draw the graph of $z = e^{-x^2-y^2}$ over the square [-2, 2] x [-2, 2] as follows:

```
xx = -2:.2:2 ;
yy = xx ;
[x, y] = meshgrid(xx, yy) ;
z = exp(-x.^2 - y.^2) ;
mesh(z)
```

79

Try this plot with surf instead of mesh. Note that you must use x.^2 and y.^2 instead of x^2 and y^2 to ensure that the function acts entry-wise on x and y.

13.3 Parametrically defined surfaces

Plots of parametrically defined surfaces can also be made. See the MATLAB functions sphere and cylinder for example. The next example displays the cover of this book, with lighting, color, and viewpoint defined in Section 13.6. First, start a figure and set up the mesh:

```
figure(1) ; clf
t = linspace(0, 2*pi, 512) ;
[u,v] = meshgrid(t) ;
```

Next, define the surface:[2]

```
a = -0.2 ; b = .5 ; c = .1 ;
n = 2 ;
x = (a*(1-v/(2*pi)).*(1+cos(u)) + c) ...
    .* cos(n*v) ;
y = (a*(1-v/(2*pi)).*(1+cos(u)) + c) ...
    .* sin(n*v) ;
z = b*v/(2*pi) + ...
    a*(1-v/(2*pi)) .* sin(u) ;
```

Plot the surface, using y to define the color, and turn off the mesh lines on the surface:

```
surf(x,y,z,y)
shading interp
```

Also try a=-0.5, which gives the back cover.

---

[2] von Seggern, CRC Standard Curves and Surfaces, 2nd ed., CRC Press, 1993, pp. 306-307.

Other three-dimensional plotting functions you may wish to explore via `help` or `doc` are `meshz`, `surfc`, `surfl`, `contour`, and `pcolor`. For plotting symbolically defined parametric surfaces (including the same seashell you plotted above), see Section 16.7.

13.4 Volume and vector visualization

MATLAB has an extensive suite of volume and vector visualization tools. The following example evaluates a function of three variables, $v=f(x,y,z)$, that represents a fluid flow problem. It returns both `v` and the coordinates (`x`, `y`, and `z`) at which the function was evaluated.

```
[x,y,z,v] = flow ;
```

Now try visualizing it. The first method plots the surface at which `v` is -3; the second plots slices of the data:

```
figure(1) ; clf
isosurface(x, y, z, v, -3)
figure(2) ; clf
slice(x, y, z, v, [3 8], 0, 0)
```

Type `doc specgraph` for more volume and vector visualization tools.

13.5 Color shading and color profile

The color shading of surfaces is set by the `shading` command. There are three settings for shading: `faceted` (default), `interpolated`, and `flat`. These are set by the commands:

```
shading faceted
shading interp
shading flat
```

Note that on surfaces produced by surf, the settings interpolated and flat remove the superimposed mesh lines. Experiment with various shadings on the surface produced above. The command shading (as well as colormap and view described below) should be entered after the surf command.

The color profile of a surface is controlled by the colormap command. Available predefined color maps include hsv (the default), hot, cool, jet, pink, copper, flag, gray, bone, prism, and white. The command colormap(cool), for example, sets a certain color profile for the current figure. Experiment with various color maps on the surface produced above. See also help colorbar.

13.6 Perspective of view

The Figure window provides a wide range of controls for viewing the figure. Select View ▶ Camera Toolbar to see these controls, or pull down the Tools menu. Try, for example, selecting Tools ▶ Rotate 3D, and then click the mouse in the Figure window and drag it to rotate the object. Some of these options can be controlled by the view and rotate3d commands, respectively.

The MATLAB function peaks generates an interesting surface on which to experiment with shading, colormap, and view. Type peaks, select Tools ▶ Rotate 3D, and click and drag the figure to rotate it.

In MATLAB, light sources and camera position can be set. Taking the peaks surface from the example above, select Insert ▶ Light, or type light to add a light source. See the online document *Using MATLAB Graphics* for camera and lighting help.

This example defines the color, shading, lighting, surface material, and viewpoint for the cover of the book:

```
axis off
axis equal
colormap(hsv(1024))
shading interp
material shiny
lighting gouraud
lightangle(80, -40)
lightangle(-90, 60)
view([-150 10])
```

14. Advanced Graphics

MATLAB possesses a number of other advanced graphics capabilities. Significant ones are bitmapped images, object-based graphics, called Handle Graphics®, and Graphical User Interface (GUI) tools.

14.1 Handle Graphics

Beyond those just described, MATLAB's graphics system provides low-level functions that let you control virtually all aspects of the graphics environment to produce sophisticated plots. The commands `set` and `get` allow access to all the properties of your plots. Try `set(gcf)` to see some of the properties of a figure that you can control. `set(gca)` lists the properties of the current axes (see Section 14.3 for an example). This system is called Handle Graphics. See *Using MATLAB Graphics* for more information.

14.2 Graphical user interface

MATLAB's graphics system also provides the ability to add sliders, push-buttons, menus, and other user interface controls to your own figures. For information on creating user interface controls, try doc uicontrol. This allows you to create interactive graphical-based applications. Try guide (short for Graphic User Interface Development Environment). This brings up MATLAB's Layout Editor window that you can use to interactively design a graphic user interface. Also see the online document *Creating Graphical User Interfaces*.

14.3 Images

The image function plots a matrix, where each entry in the matrix defines the color of a single pixel or block of pixels in the figure. image(K) paints the (i,j)th block of the figure with color K(i,j) taken from the colormap. Here is an example of the Mandelbrot set. The bottom left corner is defined as (x0,y0), and the upper right corner is (x0+d,y0+d). Try changing x0, y0, and d to explore other regions of the set (x0=-.38, y0=.64, d=.03 is also very pretty). This is also a good example of one-dimensional indexing:

```
x0 = -2 ; y0 = -1.5 ; d = 3 ; n = 512 ;
maxit = 256 ;

x = linspace(x0, x0+d, n) ;
y = linspace(y0, y0+d, n) ;
[x,y] = meshgrid(x, y) ;
C = x + y*1i ;
Z = C ;
K = ones(n, n) ;
for k = 1:maxit
    a = find((real(Z).^2 + imag(Z).^2) < 4);
    Z(a) = (Z(a)).^2 + C(a) ;
    K(a) = k ;
```

```
end
figure(1) ; clf
colormap(jet(maxit)) ;
image(x0 + [0 d], y0 + [0 d], K) ;
set(gca, 'YDir', 'normal') ;
axis equal
axis tight
```

image, by default, reverses the *y* direction and plots the
K(1,1) entry at the top left of the figure (just like the
spy function described in Section 15.5). The set
function resets this to the normal direction, so that
K(1,1) is plotted in the bottom left corner.

Try replacing the fourth argument in surf, for the
seashell example, with K, to paint the seashell surface
with the Mandelbrot set.

15. Sparse Matrix Computations

A sparse matrix is one with mostly zero entries.
MATLAB provides the capability to take advantage of
the sparsity of matrices.

15.1 Storage modes

MATLAB has two storage modes, full and sparse, with
full the default. Currently, only double or logical
vectors or two-dimensional arrays can be stored in the
sparse mode. The functions full and sparse convert
between the two modes. Nearly all MATLAB operators
and functions operate seamlessly on both full and sparse
matrices. For a matrix A, full or sparse, nnz(A) returns
the number of nonzero elements in A. An m-by-n sparse
matrix is stored in three or four one-dimensional arrays.
For a real sparse matrix, numerical values and their row
indices are stored in two arrays of size nnzmax(A) each,
but only the first nnz(A) entries are used (complex

matrices use three arrays). All entries in any given column are stored contiguously and in sorted order. A third array of size n+1 holds the positions in the other two arrays of the first nonzero entry in each column. Thus, if A is sparse, then x=A(9,:) takes much more time than x=A(:,9), and s=A(4,5) is also slow. To get high performance when dealing with sparse matrices, use matrix expressions instead of for loops and vector or scalar expressions. If you must operate on the rows of a sparse matrix A, work with the columns of A' instead.

If a full tridiagonal matrix F is created via, say,

```
F = floor(10 * rand(6))
F = triu(tril(F,1), -1)
```

then the statement S=sparse(F) will convert F to sparse mode. Try it. Note that the output lists the nonzero entries in column major order along with their row and column indices because of how sparse matrices are stored. The statement F=full(S) returns F in full storage mode. You can check the storage mode of a matrix A with the command issparse(A).

15.2 Generating sparse matrices

A sparse matrix is usually generated directly rather than by applying the function sparse to a full matrix. A sparse banded matrix can be easily created via the function spdiags by specifying diagonals. For example, a familiar sparse tridiagonal matrix is created by:

```
m = 6 ;
n = 6 ;
e = ones(n,1) ;
d = -2*e ;
T = spdiags([e d e], [-1 0 1], m, n)
```

Try it. The integral vector [-1 0 1] specifies in which diagonals the columns of [e d e] should be placed (use full(T) to see the full matrix T and spy(T) to view T graphically). Experiment with other values of m and n and, say, [-3 0 2] instead of [-1 0 1]. See doc spdiags for further features of spdiags.

The sparse analogs of eye, zeros, and rand for full matrices are, respectively, speye, sparse, and sprand. The spones and sprand functions take a matrix argument and replace only the nonzero entries with ones and uniformly distributed random numbers, respectively. sparse(m,n) creates a sparse zero matrix. sprand also permits the sparsity structure to be randomized. This is a useful method for generating simple sparse test matrices, but be careful. Random sparse matrices are not truly "sparse" because they experience catastrophic fill-in when factorized. Sparse matrices arising in real applications typically do not share this characteristic.[3]

The versatile function sparse also permits creation of a sparse matrix via listing its nonzero entries:

```
i = [1 2 3 4 4 4] ;
j = [1 2 3 1 2 3] ;
s = [5 6 7 8 9 10] ;
S = sparse(i, j, s, 4, 3)
full(S)
```

The last two arguments to sparse in the example above are optional. They tell sparse the dimensions of the matrix; if not present, then S will be max(i) by max(j). If there are repeated entries in [i j], then the entries are

---

[3] http://www.cise.ufl.edu/research/sparse/matrices.

added together. The commands below create a matrix
whose diagonal entries are 2, 1, and 1.

```
i = [1 2 3 1] ;
j = [1 2 3 1] ;
s = [1 1 1 1] ;
S = sparse(i, j, s)
full(S)
```

The entries in i, j, and s can be in any order (the same
order for all three arrays, of course), but sparse(i,j,s)
is faster if the entries are sorted in column-major order
(ascending column index j, and entries in each column
with ascending row index i) and with no duplicate
entries. In general, if the vector s lists the nonzero entries
of S and the integral vectors i and j list their
corresponding row and column indices, then
sparse(i,j,s,m,n) will create the desired sparse m-
by-n matrix S. As another example try:

```
n = 6 ;
e = floor(10 * rand(n-1,1)) ;
E = sparse(2:n, 1:n-1, e, n, n)
```

Creating a sparse matrix by assigning values to it one at a
time is exceedingly slow; ***never*** do it. The next example
constructs the same matrix as A=sparse(i,j,s,m,n)
(except for handling duplicate entries), but it should never
be used:

```
A = sparse(m,n) ;
for k = 1:length(s)
    A(i(k),j(k)) = s(k) ;
end
```

15.3 Computation with sparse matrices

The arithmetic operations and most MATLAB functions can be applied independent of storage mode. The storage mode of the result depends on the storage mode of the operands or input arguments. Operations on full matrices always give full results. If F is a full matrix, S and Z are sparse matrices, and n is a scalar, then these operations give sparse results:

```
S+S        S*S        S.*S       S.*F
S-S        S^n        S.^n       S\Z
-S         S'         S.'        S/Z
inv(S)     chol(S)    lu(S)
diag(S)    max(S)     sum(S)
```

These give full results:

```
S+F        F\S        S/F
S*F        S\F        F/S
```

except if F is a scalar, S*F, F\S, and S/F are sparse.

A matrix built from blocks, such as [A, B; C, D], is stored in sparse mode if any constituent block is sparse. To compute the eigenvalues or singular values of a sparse matrix S, you must convert S to a full matrix and then use eig or svd, as eig(full(S)) or svd(full(S)). If S is a large sparse matrix and you wish only to compute some of the eigenvalues or singular values, then you can use the eigs or svds functions (eigs(S) or svds(S)).

15.4 Ordering methods

When MATLAB solves a sparse linear system (x=A\b), it typically starts by computing the LU, QR, or Cholesky factorization of A. This usually leads to fill-in, or the creation of new nonzeros in the factors that do not appear in A. MATLAB provides several methods that attempt to reduce fill-in by reordering the rows and columns of A, Finding the best ordering is impossible in general, so fast non-optimal heuristics are used:

```
q=colamd(A)      column approximate min. degree
q=colperm(A)     sort columns by number of nonzeros
p=symamd(A)      symmetric approximate min. degree
p=symrcm(A)      reverse Cuthill-McKee
[L,U,P,Q]=lu(A)  UMFPACK's internal ordering
```

The first two find a column ordering of A and are best used for lu or qr of A(:,q). The next two are primarily used for chol(A(p,p)). Each method returns a permutation vector. The sparse lu function[4] can find its own sparsity-preserving orderings, returning them as permutation matrices P and Q (where L*U=P*A*Q). Its ordering method is based on colamd, but it also permutes P for both sparsity and numerical robustness. Try this example west0479, a chemical engineering matrix:

```
load west0479
A = west0479 ;
spy(A)
[L,U,P] = lu(A) ;
spy(L|U)
[L,U,P] = lu(A(:,colperm(A))) ;
spy(L|U)
[L,U,P] = lu(A(:,colamd(A))) ;
spy(L|U)
[L,U,P,Q] = lu(A). ;
spy(L|U)
```

---

[4] http://www.cise.ufl.edu/research/sparse/umfpack. MATLAB 7.0 uses UMFPACK 4.0. UMFPACK 4.3 includes multiple ordering strategies and selects among them automatically.

15.5 Visualizing matrices

The `spy` function introduced in the last section plots the nonzero pattern of a sparse matrix. `spy` can also be used on full matrices. It is useful for matrix expressions coming from relational operators. Try this example (see Chapter 7 for the `ddom` function):

```
A = [
-1  2  3 -4
 0  2 -1  0
 1  2  9  1
-3  4  1  1]
C = ddom(A)
figure(2)
spy(A ~= C)
spy(A > 2)
```

What you see is a picture of where A and C differ, and another picture of which entries of A are greater than 2.

16. The Symbolic Math Toolbox

The Symbolic Math Toolbox, which utilizes the Maple® kernel as its computer algebra engine, lets you perform symbolic computation from within MATLAB. Under this configuration, MATLAB's numeric and graphic environment is merged with Maple's symbolic computation capabilities. The toolbox M-files that access these symbolic capabilities have names and syntax that will be natural for the MATLAB user. Key features of the Symbolic Math Toolbox are included in the Student Version of MATLAB. Since the Symbolic Math Toolbox is not part of the Professional Version of MATLAB (by default), it may not be installed on your system, in which case this chapter will not apply.

Many of the functions in the Symbolic Math Toolbox have the same names as their numeric counterparts. MATLAB selects the correct one depending on the type of inputs to the function. Typing `doc eig` and `doc symbolic/eig` displays the help for the numeric eigenvalue function and its symbolic counterpart, respectively.

16.1 Symbolic variables

You can declare a variable as symbolic with the `syms` statement. For example,

```
syms x
```

creates a symbolic variable `x`. The statement:

```
syms x real
```

declares to Maple that `x` is a symbolic variable with no imaginary part. Maple has its own workspace. The statements `clear` or `clear x` do not undo this declaration, because it clears MATLAB's variable `x` but not Maple's variable `s`. Use `syms x unreal`, which declares to Maple that `x` may now have a nonzero imaginary part. The `clear all` statement clears all variables in both MATLAB and Maple, and thus also resets the `real` or `unreal` status of `x`. You can also assert to Maple that `x` is always positive, with `syms x positive`.

Symbolic variables can be constructed from existing numeric variables using the `sym` function. Try:

```
z = 1/10
a = sym(z)
```

```
y = rand(1)
b = sym(y, 'd')
```

although better ways to create a include:

```
a = sym('1/10')
a = 1 / sym(10)
```

If you want to ensure a precise symbolic expression, you must avoid numeric computations. Compare these three expressions. The first is only accurate to MATLAB's double-precision numeric computation (about 16 digits). The second and third avoid numeric computation completely.

```
sym(log(2))
sym('log(2)')
log(sym(2))
```

You can create a symbolic abstract function. This example declares f(x) as some unknown function of x:

```
syms x
f = sym('f(x)')
```

The syms command and sym function have many more options. See doc syms and doc sym.

16.2 Calculus

The function diff computes the symbolic derivative of a function defined by a symbolic expression. First, to define a symbolic expression, you should create symbolic variables and then proceed to build an expression as you would mathematically. For example,

```
syms x
f = x^2 * exp(x)
diff(f)
```

creates a symbolic variable x, builds the symbolic
expression $f = x^2 e^x$, and returns the symbolic derivative of
f with respect to x: `2*x*exp(x)+x^2*exp(x)` in
MATLAB notation. Try it. Next,

```
syms t
diff(sin(pi*t))
```

returns the derivative of $\sin(\pi t)$, as a function of t.

Here are examples of taking the derivative of an abstract
function, illustrating the product, quotient, and reciprocal
rules of calculus, and a special case of the chain rule. The
function `pretty` displays a symbolic expression in an
easier-to-read form resembling typeset mathematics. See
Section 16.5 for `simple`.

```
syms x n
f = sym('f(x)')
g = sym('g(x)')
pretty(diff(f*g))
pretty(diff(f/g))
pretty(diff(1/f))
pretty(simple(diff(f^n)))
```

Formats in addition to pretty include `latex`, `ccode`, and
`fortran`. Try, for example,

```
syms x a b
f = x/(a*x+b)
pretty(f)
g = int(f)
pretty(g)
latex(g)
ccode(g)
```

```
fortran(g)
int(g)
pretty(ans)
```

Partial derivatives can also be computed. Try:

```
syms x y
g = x*y + x^2
diff(g)          % computes ∂g/∂x
diff(g, x)       % also ∂g/∂x
diff(g, y)       % ∂g/∂y
```

To permit omission of the second argument for functions such as the above, MATLAB chooses a default symbolic variable for the symbolic expression. The findsym function returns MATLAB's choice. Its rule is, roughly, to choose that lower case letter, other than i and j, nearest x in the alphabet. The status of a variable (real, unreal, positive) affects its order in the list returned by findsym. You can, of course, override the default choice as shown above. Try, for example,

```
syms x x1 x2 theta
F = x * (x1*x2 + x1 - 2)
findsym(F,1)
diff(F, x)       % ∂F/∂x
diff(F, x1)      % ∂F/∂x1
diff(F, x2)      % ∂F/∂x2
G = cos(theta*x)
diff(G, theta)   % ∂G/∂theta
```

diff can compute second or higher-order derivatives. The second derivative of sin(2x) is given by either of the following two examples:

```
diff(sin(2*x), 2)
diff(sin(2*x), x, 2)
```

With a numeric argument, diff is the difference operator of basic MATLAB, which can be used to numerically approximate the derivative of a function. See doc diff or help diff for the numeric function, and doc symbolic/diff or help sym/diff for the symbolic derivative function.

The function int attempts to compute the indefinite integral (antiderivative) of a function defined by a symbolic expression. Try, for example,

```
syms a b t x y z theta
int(sin(a*t + b))
int(sin(a*theta + b), theta)
int(x*y^2 + y*z, y)
int(x^2 * sin(x))
```

Note that, as with diff, when the second argument of int is omitted, the default symbolic variable (as selected by findsym) is chosen as the variable of integration.

In some instances, int will be unable to give a result in terms of elementary functions. Consider, for example,

```
int(exp(-x^2))
int(sqrt(1 + x^3))
```

In the first case the result is given in terms of the error function erf, whereas in the second, the result is given in terms of EllipticF, a function defined by an integral.

Here is a basic integral rule with an abstract function:

```
f = sym('f(x)')
int(diff(f) / f)
```

Definite integrals can also be computed by using additional input arguments. Try, for example,

```
int(sin(x), 0, pi)
int(sin(theta), theta, 0, pi)
```

In the first case, the default symbolic variable x was used as the variable of integration to compute:

$$\int_0^{\pi} \sin\ xdx$$

whereas in the second theta was chosen. Other definite integrals you can try are:

```
int(x^5, 1, 2)
int(log(x), 1, 4)
int(x * exp(x), 0, 2)
int(exp(-x^2), 0, inf)
```

It is important to realize that the results returned are symbolic expressions, not numeric ones. The function double will convert these into MATLAB floating-point numbers, if desired. For example, the result returned by the first integral above is 21/2. Entering double(ans) then returns the MATLAB numeric result 10.5000.

Alternatively, you can use the function vpa (variable precision arithmetic; see Section 16.3) to convert the expression into a symbolic number of arbitrary precision. For example,

```
int(exp(-x^2), 0, inf)
```

gives the result:

```
1/2*pi^(1/2)
```

Then the statement:

```
vpa(ans, 25)
```

symbolically gives the result to 25 significant digits:

```
.8862269254527580136490835
```

You may wish to contrast these techniques with the MATLAB numerical integration functions `quad` and `quadl` (see Section 17.4).

The `limit` function is used to compute the symbolic limits of various expressions. For example,

```
syms h n x
limit((1 + x/n)^n, n, inf)
```

computes the limit of $(1 + x/n)^n$ as $n \to \infty$. You should also try:

```
limit(sin(x), x, 0)
limit((sin(x+h)-sin(x))/h, h, 0)
```

The `taylor` function computes the Maclaurin and Taylor series of symbolic expressions. For example,

```
taylor(cos(x) + sin(x))
```

returns the fifth order Maclaurin polynomial approximating $\cos(x) + \sin(x)$. This returns the eighth degree Taylor approximation to $\cos(x^2)$ centered at the point $x_0 = \pi$:

```
taylor(cos(x^2), 8, x, pi)
```

16.3 Variable precision arithmetic

Three kinds of arithmetic operations are available:

numeric	MATLAB's floating-point arithmetic
rational	Maple's exact symbolic arithmetic
VPA	Maple's variable precision arithmetic

One can obtain exact rational results with, for example,

```
s = simple(sym('13/17 + 17/23'))
```

You are already familiar with numeric computations. For example, with format long,

```
pi*log(2)
```

gives the numeric result 2.17758609030360.

MATLAB's numeric computations are done in approximately 16 decimal digit floating-point arithmetic. With vpa, you can obtain results to arbitrary precision, within the limitations of time and memory. Try:

```
vpa('pi * log(2)')
vpa(sym(pi) * log(sym(2)))
vpa('pi * log(2)', 50)
```

The default precision for vpa is 32. Hence, the two results are accurate to 32 digits, whereas the third is accurate to the specified 50 digits. Ludolf van Ceulen (1540-1610) calculated π to 36 digits. The Symbolic Math Toolbox can quite easily compute π to 10,000 digits or more. Try:

```
pretty(vpa('pi', 10000))
```

The default precision can be changed with the function
digits. While the rational and VPA computations can
be more accurate, they are in general slower than numeric
computations. If you pass a numeric expression to vpa,
MATLAB will evaluate it numerically first, so use a
symbolic expression or place the expression in quotes.
Compare your results, above, with:

```
vpa(pi * log(2))
```

which is accurate to only about 16 digits (even though 32
digits are displayed). This is a common mistake with the
use of vpa and the Symbolic Math Toolbox in general.

16.4 Numeric and symbolic subsitution

Once you have a symbolic expression, you can modify it
or evaluate it numerically with the subs function. The
function subs replaces all occurrences of the symbolic
variable in an expression by a specified second
expression. This corresponds to composition of two
functions. Try, for example,

```
syms x s t
subs(sin(x), x, pi/3)
subs(sin(x), x, sym(pi)/3)
double(ans)
subs(g*t^2/2, t, sqrt(2*s))
subs(sqrt(1-x^2), x, cos(x))
subs(sqrt(1-x^2), 1-x^2, cos(x))
```

The general idea is that in the statement
subs(expr,old,new) the third argument (new)
replaces the second argument (old) in the first argument
(expr). Compare the first two examples above. The
result is numeric if all variables in the expression are
substituted with numeric values, or symbolic otherwise.

You can substitute multiple symbolic expressions, numeric expressions, or any combination, using cell arrays of symbolic or numeric values. Try:

```
syms x y
S = x^y
subs(S, x, 3)
subs(S, {x y}, {3 2})
subs(S, {x y}, {3 x+1})
```

You perform multiple substitutions for any one symbolic variable, which returns a matrix of symbolic expressions or numeric values. Try this, which constructs a function F, finds its derivative G, and evaluates G at x=0:.1:1.

```
syms x
F = x^2 * sin(x)
G = diff(F)
subs(G, x, 0:.1:1)
```

Also try:

```
a = subs(S, y, 1:9)
a(3)
a = subs(S, {x y},{2*ones(9,1) (1:9)'})
```

The first expression returns a row vector containing the symbolic expressions x, x^2, ... x^9. The second substitution returns a numeric column vector containing the powers of 2 from 2 to 512. Each entry in the cell array must be of the same size.

Substitution acts just like composition in calculus. Taking the derivative of function composition illustrates the chain rule of calculus:

```
f = sym('f(x)')
g = sym('g(x)')
```

```
diff(subs(f, g))
pretty(ans)
```

16.5 Algebraic simplification

Convenient algebraic manipulations of symbolic
expressions are available.

The function expand distributes products over sums and
applies other identities, whereas factor attempts to do
the reverse. The function collect views a symbolic
expression as a polynomial in its symbolic variable
(which may be specified) and collects all terms with the
same power of the variable. To explore these capabilities,
try the following:

```
syms a b x y z
expand((a + b)^5)
factor(ans)
expand(exp(x + y))
expand(sin(x + 2*y))
factor(x^6 - 1)
collect(x * (x * (x + 3) + 5) + 1)
horner(ans)
collect((x + y + z)*(x - y - z))
collect((x + y + z)*(x - y - z), y)
collect((x + y + z)*(x - y - z), z)
diff(x^3 * exp(x))
factor(ans)
```

The powerful function simplify applies many identities
in an attempt to reduce a symbolic expression to a simple
form. Try, for example,

```
simplify(sin(x)^2 + cos(x)^2)
simplify(exp(5*log(x) + 1))
d = diff((x^2 + 1)/(x^2 - 1))
simplify(d)
```

The alternate function `simple` computes several simplifications and chooses the shortest of them. It often gives better results on expressions involving trigonometric functions. Try the following commands:

```
simplify(cos(x) + (-sin(x)^2)^(1/2))
simple   (cos(x) + (-sin(x)^2)^(1/2))
simplify((1/x^3+6/x^2+12/x+8)^(1/3))
simple   ((1/x^3+6/x^2+12/x+8)^(1/3))
```

The function `factor` can also be applied to an integer argument to compute the prime factorization of the integer. Try, for example,

```
factor(sym('4248'))
factor(sym('4549319348693'))
factor(sym('4549319348597'))
```

16.6 Two-dimensional graphs

The MATLAB function `fplot` (see Section 12.3) provides a tool to conveniently plot the graph of a function. Since it is, however, the name or handle of the function to be plotted that is passed to `fplot`, the function must first be defined in an M-file (or else be a built-in function or anonymous function).

In the Symbolic Math Toolbox, `ezplot` lets you plot the graph of a function directly from its defining symbolic expression. For example, to plot a function of one variable try:

```
syms t x y
ezplot(sin(2*x))
ezplot(t + 3*sin(t))
ezplot(2*x/(x^2 - 1))
ezplot(1/(1 + 30*exp(-x)))
```

By default, the *x*-domain is [-2*pi, 2*pi]. This can be overridden by a second input variable, as with:

```
ezplot(x*sin(1/x), [-.2 .2])
```

You will often need to specify the *x*-domain and *y*-domain to zoom in on the relevant portion of the graph. Compare, for example,

```
ezplot(x*exp(-x))
ezplot(x*exp(-x), [-1 4])
```

ezplot attempts to make a reasonable choice for the *y*-axis. With the last figure, select Edit ▶ Axes Properties in the Figure window and modify the *y*-axis to start at -3, and hit enter. Changing the *x*-axis in the Property Editor does not cause the function to be reevaluated, however.

To plot an implicitly defined function of two variables:

```
ezplot(x^2 + y^2 - 1)
```

which plots the unit circle over the default *x*-domain and *y*-domain of [-2*pi, 2*pi]. Since this is too large for the unit circle, try this instead:

```
ezplot(x^2 + y^2 - 1, [-1 1 -1 1])
```

The first two entries in the second argument define the *x*-domain. The second two define the *y*-domain. If the *y*-domain is the same as the *x*-domain, then you only need to specify the *x*-domain (see the next example).

In both of the previous examples, you plotted a circle but it looks like an ellipse. This is because with auto-scaling, the *x* and *y* axes are not equal. Fix this by typing:

```
axis equal
```

To plot a parameterized function, provide two function arguments. Try this, which plots a cycloid over the domain -4π to 4π.

```
x = t-sin(t)
y = 1-cos(t)
ezplot(x,y, [-4*pi 4*pi])
```

The `ezpolar` function creates polar plots. Try creating a three-leaf rose and a hyperbolic spiral:

```
ezpolar(sin(3*t))
ezpolar(1/t, [1 10*pi])
```

Entering the command `funtool` (no input arguments) brings up three graphic figures, two of which will display graphs of functions and one containing a control panel. This function calculator lets you manipulate functions and their graphs for pedagogical demonstrations. Type `doc funtool` for details.

16.7 Three-dimensional surface graphs

MATLAB has several easy-to-use functions for creating three-dimensional surface graphs.

ezcontour	3-D contour plot
ezcontourf	3-D filled contour plot
ezmesh	3-D mesh plot
ezmeshc	3-D mesh and contour plot

ezsurf	3-D surface plot
ezsurfc	3-D surface and contour plot

Here is an interesting function to try:

```
syms x y
f = sin((x^2+y)/2)/(x^2-x+2)
ezsurfc(f)
```

Try each of these plotting functions with this function f. For this function, ezcontourf gives more information than ezcontour because the function fluctuates across a single contour in several regions. The default domain for *x* and *y* is -2π to 2π. You can change this with an optional second parameter. Try:

```
ezsurf(f, [-4 4 -pi pi])
```

which defines the *x*-domain as -4 to 4, and the *y*-domain as -π to π. The appearance of the plots can be modified by the shading command after the figure is plotted (see Section 13.5).

Functions with discontinuities or singularities can cause difficulty for these graphing functions. Here is an example that is similar to the function f above,

```
f = sin(abs(sqrt(x^2+y)))/(x^2-x+2)
ezsurf(f)
```

Click the rotate button

in the figure window, then click and drag the graph itself. The function touches the *z=0* plane along the curve

defined by $y = -x^2$, but the graph does not capture this
property very well because the gradient is not defined
along that curve. To plot this function accurately, you
would need to define your own mesh points, compute the
function numerically, and use surf or another numerical
graphing function instead.

The four mesh and surface functions listed above can also
plot parameterized surface functions. The first three
arguments are the $x(s,t)$, $y(s,t)$, and $z(s,t)$ functions, and
the last (optional) argument defines the domain. To
create a symbolic seashell, start a new figure and define
your symbolic variables:

```
figure(1) ; clf
syms u v x y z
```

Next, define x, y, and z, just as you did for the numeric
seashell in Section 13.3. The MATLAB statements are
the same, except that now these variables are defined
symbolically, not numerically. Plot the symbolic surface:

```
ezsurfc(x,y,z,[0 2*pi])
```

Turn off the axis and set the shading, material, lighting,
and viewpoint, just as you did in Section 13.3 and 13.6.
You cannot change the ezsurfc color.

16.8 Three-dimensional curves

Parameterized 3-D curves are plotted with ezplot3. Try
this example, which combines a folium of Descartes in
the x-y plane with a sinusoid in the z direction:

```
syms x y z t
x = 3*t / (1+t^3)
y = 3*t^2 / (1+t^3)
```

```
z = sin(t)
ezplot3(x,y,z)
```

The default domain for t is 0 to 2π. Here is an example of how to change it:

```
ezplot3(x,y,z,[-.9 10])
```

The ezplot3 function can animate the plot so that you can observe how x, y, and z depend on t. Try both of these examples. The ball moves quickly over the first half of the curve but more slowly over the second half:

```
ezplot3(x,y,z,'animate')
ezplot3(x,y,z, [-.9 10], 'animate')
```

The 2-D curve plotting function ezplot cannot animate its plot, but you can do the same with ezplot3. Just give it a z argument of zero. Try:

```
syms z
z = 0
ezplot3(x,y,z,'animate')
```

and then rotate the graph so that you are viewing the *x-y* plane. Click the rotate button and drag the graph, or right-click the graph and select Go to X-Y view. Then click the Repeat button in the bottom left corner.

16.9 Symbolic matrix operations

This toolbox lets you represent matrices in symbolic form as well as MATLAB's numeric form. Given numeric matrix a, sym(a) converts a to a symbolic matrix. Try:

```
a = magic(3)
A = sym(a)
```

The function double(A) converts the symbolic matrix back to a numeric one.

Symbolic matrices can also be generated. Try, for example,

```
syms a b s
K = [a + b, a - b ; b - a, a + b]
G = [cos(s), sin(s); -sin(s), cos(s)]
```

Here G is a symbolic Givens rotation matrix.

Algebraic matrix operations with symbolic matrices are computed as you would in MATLAB:

K+G	matrix addition
K-G	matrix subtraction
K*G	matrix multiplication
inv(G)	matrix inversion
K\G	left matrix division
K/G	right matrix division
G^2	power
G.'	transpose
G'	conjugate transpose (Hermitian)

These operations are illustrated by the following, which use the matrices K and G generated above. The last expression demonstrates that G is orthogonal.

```
L = K^2
collect(L)
factor(L)
diff(L, a)
int(K, a)
J = K/G
simplify(J*G)
simplify(G*(G.'))
```

The initial result of the basic operations may not be in the form desired for your application; so it may require further processing with simplify, collect, factor, or expand. These functions, as well as diff and int, act entry-wise on a symbolic matrix.

16.10 Symbolic linear algebraic functions

The primary symbolic matrix functions are:

det	determinant
.'	transpose
'	Hermitian (conjugate transpose)
inv	inverse
null	basis for nullspace
colspace	basis for column space
eig	eigenvalues and eigenvectors
poly	characteristic polynomial
svd	singular value decomposition
jordan	Jordan canonical form

These functions will take either symbolic or numeric arguments. Computations with symbolic rational matrices can be carried out exactly. Try, for example,

```
c = floor(10*rand(4))
D = sym(c)
A = inv(D)
inv(A)
inv(A) * A
det(A)
b = ones(1,4)
x = b/A
x*A
A^3
```

These functions can, of course, be applied to general
symbolic matrices. For the matrices K and G defined in
the previous section, try:

```
inv(K)
simplify(inv(G))
p = poly(G)
simplify(p)
factor(p)
X = solve(p)
for j = 1:4
    X = simple(X)
end
pretty(X)
e = eig(G)
for j = 1:4
    e = simple(e)
end
pretty(e)
y = svd(G)
for j = 1:4
    y = simple(y)
end
pretty(y)
syms s real
r = svd(G)
r = simple(r)
pretty(r)
syms s unreal
```

The simple function had to be repeated several times for
some of the examples to get the simplest possible result.

Compare y and r. If you do not declare s as real, the
svd of the 2-by-2 Givens rotation matrix does not
demonstrate that the singular values are all equal to one.

A typical exercise in a linear algebra course is to
determine those values of t so that, say,

```
A = [t 1 0 ; 1 t 1 ; 0 1 t]
```

is singular. The following simple computation:

```
syms t
A = [t 1 0 ; 1 t 1 ; 0 1 t]
p = det(A)
solve(p)
```

shows that this occurs for t = 0, √2, and √–2. See Section 16.11 for the `solve` function.

The function `eig` attempts to compute the eigenvalues and eigenvectors in an exact closed form. Try, for example,

```
for n = 4:6
    A = sym(magic(n))
    [V, D] = eig(A)
end
```

Except in special cases, however, the result is usually too complicated to be useful. Try, for example, executing:

```
A = sym(floor(10 * rand(3)))
[V, D] = eig(A)
pretty(V)
```

a few times. The eigenvectors V are not very pretty. For this reason, it is usually more efficient to do the computation in variable-precision arithmetic, as is illustrated by:

```
A = vpa(floor(10 * rand(3)))
[V, D] = eig(A)
```

The comments above regarding `eig` apply as well to the computation of the singular values of a matrix by `svd`, as can be observed by repeating some of the computations above using `svd` instead of `eig`.

16.11 Solving algebraic equations

For a symbolic expression S, the statement `solve(S)` will attempt to find the values of the symbolic variable for which the symbolic expression is zero. The `solve` function cannot solve all equations. It does well with polynomial equations, but can have difficulty with trigonometric or other transcendental equations. If an exact symbolic solution is indeed found, you can convert it to a floating-point solution, if desired. If an exact symbolic solution cannot be found, then a variable precision one is computed. Here are three similar equations. The first returns a symbolic result, the second a numeric result, and the last one fails.

```
syms x b
solve(2^x - b)
solve(2^x + 3^x - 1)
solve(2^x + 3^x - b)
```

If you have an expression that contains several symbolic variables, you can solve for a particular variable by including it as an input argument in `solve`. The default variable solved for is x, or the one closest (alphabetically) to x if x is not a variable in the equation.

Try this example; note that X contains four solutions:

```
syms x y z
f = cos(x) + tan(x)
X = solve(f)
pretty(X)
double(X)
vpa(X)
for i = 1:4
  s = simple(subs(f, x, X(i)))
end
```

Here are some more examples:

```
Y = solve(cos(x) - x)
Z = solve(x^2 + 2*x - 1)
pretty(Z)
a = solve(x^2 + y^2 + z^2 + x*y*z)
pretty(a)
b = solve(x^2 + y^2 + z^2 + x*y*z, y)
pretty(b)
```

a is a solution in the variable x, and b is a solution in y.

The inputs to solve can be quoted strings or symbolic expressions. To solve an equation whose right-hand side is not zero, use a quoted string or rearrange the equation:

```
X = solve('log(x) = x - 2')
X = solve(log(x) - x + 2)
vpa(X)
X = solve('2^x = x + 2')
X = solve(2^x - x - 2)
vpa(X)
```

This solves for the variable a:

```
solve('1 + (a+b)/(a-b) = b', 'a')
```

This solves the same for b, finding two solutions:

```
solve('1 + (a+b)/(a-b) = b', 'b')
```

The solution to the next example should be familiar. Try:

```
syms a b c x
solve(a*x^2 + b*x + c, x)
pretty(ans)
```

The function solve can also compute solutions of systems of general algebraic equations. To solve, for

114

example, the nonlinear system below, it is convenient to
first express the equations as strings.

```
S1 = 'x^2 + y^2 + z^2 = 2'
S2 = 'x + y = 1'
S3 = 'y + z = 1'
```

The solutions are then computed by:

```
[X, Y, Z] = solve(S1, S2, S3)
```

If you request the set of solutions in a single output with
multiple unknowns, a **struct** is returned. Try

```
a = solve(S1, S2, S3)
a.x
a.y
a.z
```

If you alter S2 to:

```
S2 = 'x + y + z = 1'
```

then the solution computed by:

```
[X, Y, Z] = solve(S1, S2, S3)
```

will be given in terms of square roots. If you prefer
solving symbolic expressions instead of strings, try

```
syms x y z
S1 = x^2 + y^2 + z^2 - 2
S2 = x + y - 1
S3 = y + z - 1
a = solve(S1, S2, S3)
```

The output of **solve** is in alphabetical order. For
example, if you changed the name of z to w in these three

equations the results would be returned in the order
[W,X,Y]. The solve function can take quoted strings or
symbolic expressions as input arguments, but you cannot
mix the two types of inputs.

16.12 Solving differential equations

The function dsolve solves ordinary differential
equations. The symbolic differential operator is D:

```
Y = dsolve('Dy = x^2*y', 'x')
```

produces the solution C1*exp(1/3*x^3) to the
differential equation $y' = x^2 y$. The solution to an initial
value problem can be computed by adding a second
symbolic expression giving the initial condition.

```
Y = dsolve('Dy = x^2*y', 'y(0)=4', 'x')
```

Notice that in both examples above, the final input
argument, 'x', is the independent variable of the
differential equation. If no independent variable is
supplied to dsolve, then it is assumed to be t. The
higher order symbolic differential operators D2, D3, ...
can be used to solve higher order equations. Try:

```
dsolve('D2y + y = 0')
dsolve('D2y + y = x^2', 'x')
dsolve('D2y + y = x^2', ...
    'y(0) = 4', 'Dy(0) = 1', 'x')
dsolve('D2y - Dy = 2*y')
dsolve('D2y + 6*Dy = 13*y')
dsolve('D3y - 3*Dy = 2*y')
pretty(ans)
```

Systems of differential equations can also be solved:

```
E1 = 'Dx = -2*x + y'
E2 = 'Dy = x - 2*y + z'
E3 = 'Dz = y - 2*z'
```

The solutions are then computed with:

```
[x, y, z] = dsolve(E1, E2, E3)
pretty(x)
pretty(y)
pretty(z)
```

You can explore further details with doc dsolve.

16.13 Further Maple access

The following features are not available in the Student Version of MATLAB.

Over 50 special functions of classical applied mathematics are available in the Symbolic Math Toolbox. Enter doc mfunlist to see a list of them. These functions can be accessed with the function mfun, for which you are referred to doc mfun for further details. The maple function allows you to use expressions and programming constructs in Maple's native language, which gives you full access to Maple's functionality. See doc maple, or mhelp *topic*, which displays Maple's help text for the specified topic. The Extended Symbolic Math Toolbox provides access to a number of Maple's specialized libraries of procedures. It also provides for use of Maple programming features.

17. Polynomials, Interpolation, and Integration

Polynomial functions are frequently used by numerical methods, and thus MATLAB provides several routines that operate on polynomials and piece-wise polynomials.

17.1 Representing polynomials

Polynomials are represented as vectors of their coefficients, so $f(x)=x^3-15x^2-24x+360$ is simply

```
p = [1 -15 -24 360]
```

The roots of this polynomial (15, $\sqrt{24}$, and $-\sqrt{24}$):

```
r = roots(p)
```

Given a vector of roots r, poly(r) constructs the coefficients of the polynomial with those roots. With a little bit of roundoff error, you should see the original polynomial. Try it.

The poly function also computes the characteristic polynomial of a matrix whose roots are the eigenvalues of the matrix. The polynomial $f(x)$ was chosen as the characteristic equation of the magic(3) matrix. Try:

```
A = magic(3)
s = poly(A)
roots(s)
eig(A)
f = poly(sym(A))
solve(f)
eig(sym(A))
```

17.2 Evaluating polynomials

You can evaluate a polynomial at one or more points with the polyval function.

```
x = -1:2 ;
y = polyval(p,x)
```

Compare y with x.^3-15*x.^2-24*x+360. You can construct a symbolic polynomial from the coefficient vector p and back again:

```
syms x
f = poly2sym(p)
sym2poly(f)
```

17.3 Polynomial interpolation

Polynomials are useful as easier-to-compute approximations of more complicated functions, via a Taylor series expansion or by a low-degree best-fit polynomial using the polyfit function. The statement:

```
p = polyfit(x, y, n)
```

finds the best-fit n-degree polynomial that approximates the data points x and y. Try this example:

```
x = 0:.1:pi ;
y = sin(x) ;
p = polyfit(x, y, 5)
figure(1) ; clf
ezplot(@sin, [0 pi])
hold on
xx = 0:.001:pi ;
plot(xx, polyval(p,xx), 'r-')
```

Piecewise-polynomial interpolation is typically better than a single high-degree polynomial. Try this example:

```
n = 10
x = -5:.1:5 ;
y = 1 ./ (x.^2+1) ;
p = polyfit(x, y, n)
figure(2) ; clf
ezplot(@(x) 1/(x^2+1))
hold on
xx = -5:.01:5 ;
plot(xx, polyval(p,xx), 'r-')
```

As n increases, the error in the center improves but
increases dramatically near the endpoints. The spline
and pchip functions compute piecewise-cubic
polynomials which are better for this problem. Try:

```
figure(3) ; clf
yy = spline(x, y, xx) ;
plot(xx, yy, 'g')
```

Alternatively, with two inputs, spline and pchip return
a struct that contains the piecewise polynomial, which
can be later evaluated with ppval. Try:

```
figure(4) ; clf
pp = spline(x, y)
yy = ppval(pp, xx) ;
plot(xx, yy, 'c')
```

The spline function computes the conventional cubic
spline, with a continuous second derivative. In contrast,
pchip returns a piecewise polynomial with a
discontinuous second derivative, but it preserves the
shape of the function better than spline.

Polynomial multiplication and division (convolution and
deconvolution) are performed by the conv and deconv
functions. MATLAB also has a built-in fast Fourier
transform function, fft.

```
```

17.4 Numeric integration (quadrature)

The quad and quadl functions are the numeric
equivalent of the symbolic int function, for computing a
definite integral. Both rely on polynomial
approximations of subintervals of the function being
integrated. quadl is a higher-order method that can be
more accurate. The syntax quad(@f,a,b) computes an
approximate of the definite integral,

$$\int_a^b f(x)dx$$

Compare these examples:

```
quad(@(x) x.^5, 1, 2)
quad(@log, 1, 4)
quad(@(x) x .* exp(x), 0, 2)
quad(@(x) exp(-x.^2), 0, 1e6)
quad(@(x) sqrt(1 + x.^3), -1, 2)
quad(@(x) real(airy(x)), -3, 3)
```

with the same results from the Symbolic Toolbox:

```
int('x^5', 1, 2)
int('log(x)', 1, 4)
int('x * exp(x)', 0, 2)
int('exp(-x^2)', 0, inf)
int('sqrt(1 + x^3)', -1, 2)
int('real(airy(x))', -3, 3)
```

Symbolic integration (int) can find a simple closed-form
solution to the first four examples, above. The next is not
in closed form, and the last example cannot be solved by
int at all. It can only be computed numerically, with
quad.

The function f provided to quad and quadl must operate on a vector x and return f(x) for each component of the vector. An optional fourth argument to quad and quadl modifies the error tolerance. Double and triple integrals are evaluated by dblquad and triplequad. Array-valued functions are integrated with quadv.

18. Solving Equations

Solving equations is at the core of what MATLAB does. Let us look back at what kinds of equations you have seen so far in the book. Next, in this chapter you will learn how MATLAB finds numerical solutions to nonlinear equations and systems of differential equations.

18.1 Symbolic equations

The Symbolic Toolbox can solve symbolic linear systems of equations using backslash (Section 16.9), nonlinear systems of equations using the solve function (Section 16.11), and systems of differential equations using dsolve (Section 16.12). The rest of MATLAB focuses on finding numeric solutions to equations, not symbolic.

18.2 Linear systems of equations

The pervasive and powerful backslash operator solves linear systems of equations of the form A*x=b (Sections 3.3, 15.3, and 16.9). The expression x=A\b handles the case when A is square or rectangular (under- or over-determined), full-rank or rank-deficient, full or sparse, numeric or symbolic, symmetric or unsymmetric, real or complex, and all but one reasonable combination of this extensive list (backslash does not work with complex rectangular sparse matrices). It efficiently handles triangular, permuted triangular, symmetric positive-definite, and Hessenberg matrices. Further details for the case when A is sparse are discussed in Chapter 15. When the matrix has specific known properties, the `linsolve` function can be faster (see Section 5.5, and a related Fortran code in Chapter 10).

18.3 Polynomial roots

Solving the function $f(x)=0$ for the special case when f is a polynomial and x is a scalar is discussed in Section 17.1. The more general case is discussed below.

18.4 Nonlinear equations

The `fzero` function finds a numerical solution to $f(x)=0$ when f is a real function over the real domain (both x and $f(x)$ must be real scalars). This is useful when an analytic solution is not possible. You must supply either an initial guess, or two values of x for which the function differs in sign. Here is a simple example that computes $\sqrt{2}$.

```
fzero(@(x) x^2-2, 1)
```

The `fzero` function can only find an x for which $f(x)$ crosses the x-axis. If the sign of $f(x)$ does not differ on either side of x, the zero point x will not be found. Try this example. Create two anonymous functions (regular M-files can also be used):

```
fa = @(x) (x-2)^2
fb = @(x) (x-2)^2 - 1e-12
```

The zero of `fa` cannot be found, and neither can a zero of `fb` be found if your initial guess is too far from the solution. Both of these examples will fail:

```
fzero(fa, 1)
fzero(fb, 3)
```

Both functions can be easily solved with the Symbolic
Toolbox. Note that solve correctly reports that 2 is a
double root of (x-2)^2. Try:

```
syms x
solve((x-2)^2)
s = solve((x-2)^2-1e-12)
fb(s(1))
fb(s(2))
```

The zeros of fb can be found numerically only if you
guess close enough, or if you provide two initial values of
x for which fb differs in sign:

```
fzero(fb, 2)
format long
fzero(fb, [2 3])
fzero(fb, [1 2])
```

All of the functions used in the examples so far can be
solved analytically. Here is one that cannot (also plot the
function so that you can see where it crosses the *x*-axis):

```
f = @(x) real(airy(x))
figure(1) ; clf
ezplot(f)
solve('real(airy(x))')
```

The first zero is easy to compute numerically:

```
s = fzero(f, 0)
hold on
plot(s, f(s), 'ro')
```

The `fminbnd` function finds a local minimum of a function, given a fixed interval. This example looks for a minimum in the range -4 to 0.

```
xmin = fminbnd(f, -4, 0)
plot(xmin, f(xmin), 'ko')
```

To find a local maximum, simply find the minimum of -f.

```
g = @(x) -real(airy(x))
xmax = fminbnd(g, -5, -4)
plot(xmax, f(xmax), 'ko')
```

Now find the zero between these two values of x:

```
s = fzero(f, [xmax xmin])
plot(s, f(s), 'ro')
```

The `fminbnd` function can only find minima of real-valued functions of a real scalar. To find a local minimum of a scalar function of a real vector x, use `fminsearch` instead. It takes an initial guess for x rather than an interval.

18.5 Ordinary differential equations

The symbolic solution to the ordinary differential equation $y' = t^2 y$ appears in Section 16.12. Here is the same ODE, with a specific initial value of $y(0)=1$, along with its symbolic solution.

```
syms t y
Y = dsolve('Dy = t^2*y', 'y(0)=1', 't')
```

Not all ODEs can be solved analytically, so MATLAB provides a suite of numerical methods. The primary method for initial value problems is `ode45`. For an ODE of the form $y' = f(t,y)$, the basic usage is:

```
[tt,yy] = ode45(@f, tspan, y0)
```

where @f is a handle for a function yprime=f(t,y) that
computes the derivative of y, tspan is the time span to
compute the solution (a 2-element vector), and y0 is the
initial value of y. The variable t is a scalar, but y can be
a vector. The solution is a column vector tt and a matrix
yy. At time tt(i) the numerical approximation to y is
yy(i,:).

To solve this ODE numerically, create an anonymous
function:

```
f1 = @(t,y) t^2 * y
```

Now you can compute the numeric solution:

```
[tr,yr] = ode45(f1, [0 2], 1) ;
```

Compare it with the symbolic solution:

```
ts = 0:.05:2 ;
ys = subs(Y, t, ts) ;
figure(2) ; clf
plot(ts,ys, 'r-', tr,yr, 'bx') ;
legend('symbolic', 'numeric')
ys = subs(Y, t, tr) ;
[tr ys yr ys-yr]
err = norm(ys-yr) / norm(ys)
```

To solve higher-order ODEs, you need to convert your
ODE into a first-order system of ODEs. Let us start with
the ODE $y''+y=t^2$ with initial values $y(0)=1$ and $y'(1)=0$.
The symbolic solution to this ODE appears in Section
16.12, but here is the solution with initial values
specified:

```
Y = dsolve('D2y + y = t^2', ...
      'y(0)=1', 'Dy(0)=0', 't')
```

Define $y_1 = y$ and $y_2 = y'$. The new system is $y_2' = t^2 - y_1$ and $y_1' = y_2$. Create an anonymous function:

```
f2 = @(t,y) [y(2) ; t^2-y(1)]
```

The function f2 returns a 2-element column vector. The first entry is y_1' and the second is y_2'. We can now solve this ODE numerically:

```
[tr,yy] = ode45(f2, [0 2], [1 0]') ;
yr = yy(:,1) ;
```

Note that ode45 returns a 61-by-2 solution yy. Row i of yy contains the numerical approximation to y_1 and y_2 at time tr(i). Compare the symbolic and numeric solutions using the same code for the previous ODE.

MATLAB's ode45 can return a structure s=ode45(...) which can be used by deval to evaluate the numerical solution at any time t that you specify. There are seven other ODE solvers, able to handle stiff ODEs and for differential algebraic equations. Some can be more efficient, depending on the type of ODE you are trying to solve. Type doc ode45 for more information.

18.6 Other differential equations

Delay differential equations (DDEs) are solved by dde23. The function bvp4c solves boundary value ODE problems. Finally, partial differential equations are solved with pdepe and pdeval. See the online help facility for more information on these ODE, DDE, and PDE solvers.

19. Displaying Results

The `format` command provides basic control over how your results are printed in the Command window. For example, if you want a trigonometric table with just a few digits of precision, you could do:

```
warning('off','MATLAB:divideByZero')
format short
x = [0:.1:pi]' ;
f = {@sin, @cos, @tan, @cot} ;
y = x ;
for i = 1:length(f)
    y = [y f{i}(x)] ;
end
disp(y)
```

The cell array `f` is used in the next example; otherwise a simpler way to construct `y` would be:

```
y = [x sin(x) cos(x) tan(x) cot(x)] ;
```

You can increase the number of digits printed with `format long`, but that does not allow you to define how many digits are printed. If you tried to add `pi/2` to the table, the `tan` column would contain a huge (erroneous) value causes the rest of the digits in the table to be obscured. Try adding the statement `x=[x ; pi/2]` after `x` is first defined.

This problem is where `fprintf` is useful. If you know C, it acts just like the standard C `fprintf`, except that the reference to the file is optional in the MATLAB `fprintf`, and MATLAB's `fprintf` can print arrays.

The basic syntax (like `printf` in C) is:

```
fprintf(format_string, arg1, arg2, ...)
```

The format string tells MATLAB how to print each
argument (*arg1*, *arg2*, ...). It contains plain text, which
is printed verbatim, plus special conversion codes that
start with '%' (to print an argument) or '\' (to print a
special character such as a newline, tab, or backslash).

The basic syntax for a conversion code is %*W.Pc*, where *W*
is the optional field width (the total number of characters
used to represent the number), *P* is the optional precision
(the number of digits to the right of the decimal point),
and *c* is the conversion type. Both *W* and *P* are fixed
integers. The dot before the *P* field is required only if *P* is
specified. The most common conversion types are:

d	decimal (integer)
e	exponential notation (as in 2.3e+002)
f	fixed-point notation
g	e or f, whichever is more compact
s	string

Special characters include \n for newline, \t for tab, and
\\ for backslash itself. A single quote is either \' or
two single quotes (''').

Here is a simple example that prints pi with 8 digits past
the decimal point, in a space of 12 characters:

```
fprintf('pi is %12.8f\n', pi)
```

Try changing the 12 to 14, and you will see how fprintf
pads the string for pi to make it 14 characters wide. Note
the last character is '\n', which is a newline. If this
were excluded, the next line of output would start at the

end of this line. Sometimes that is what you want (see below for an example).

Unlike `printf` or `fprintf` in the C language, MATLAB's `fprintf` can print arrays. It accesses an array column by column, and reuses the format string as needed. This simple example prints the `magic(3)` array. It also gives you an example of how to print a backslash and a single quote:

```
A = magic(3)
fprintf('%4.2f %4.2f %4.2f\n', A')
b = (1:3)' ;
fprintf('A\\b is [%g %g %g]''\n', A\b);
```

The array A is transposed in the first `fprintf`, because `fprintf` cycles through its data column by column, but each use of the format string prints a single line of text as one row of characters on the Command window. Fortunately it makes no difference for vectors:

```
fprintf('x is %d\n', 1:5)
fprintf('x is %d\n', (1:5)')
```

Here is a way of adding extra information to your display:

```
fprintf( ...
'row %d is %4.2f %4.2f %4.2f\n', ...
[(1:3)' A]')
```

Here is a revised trigonometric table using `fprintf` instead. A header has been added as well:

```
x = [0:.1:pi]' ;
f = {@sin, @cos, @tan, @cot} ;
y = x ;
fprintf('    x') ;
for i = 1:length(f)
```

```
        fprintf('    %s(x)',func2str(f{i}));
    y = [y f{i}(x)] ;
end
fprintf('\n') ;
fprintf( ...
'%3.2f %9.4f %9.4f %9.4f %9.4f\n',y);
```

`fprintf`, by default, prints to the Command window. You can instead open a file, write to it with `fprintf`, and close the file. Add:

```
fid = fopen('mytable.txt', 'w') ;
```

to the beginning of the example. Add `fid` as the first argument to each `fprintf`. Finally, close the file at the end with the statement:

```
fclose(fid) ;
```

Your table is now in the file `mytable.txt`.

The `sprintf` function is just like `fprintf`, except that it sends its output to a string instead of the Command window or a file. It is useful for plot titles and other annotation, as in:

```
title(sprintf('The result is %g', pi))
```

You cannot control the field width or precision with a variable as you can in the C `printf` or `fprintf`, but string concatenation along with `sprintf` or `num2str` can help here. Try:

```
for n = 1:16
    s = num2str(n) ;
    s = ['%2d digits: %.' s 'g\n'] ;
    fprintf(s, n, pi) ;
end
```

20. Cell Publishing

Cell publishing creates nicely formatted reports of MATLAB code, command window text output, figures, and graphics in HTML, LaTeX, XML, Microsoft Word, or Microsoft Powerpoint.

The term *cell publishing* has nothing to do with the cell array data type. In this context, a cell is a section of an M-file that corresponds to a section of your report. A cell starts with a cell divider, which is a comment with two percent signs at the beginning of a line, and ends either at the start of the next cell, or the end of the M-file. Cell publishing is normally done via scripts, not functions.

Create a new M-file, and select the Editor menu item Cell ► Enable Cell Mode. Try this 2-cell example:

```
%% Integrate a function
syms x
f = x^2
e = int(f)

%% Plot the results
figure(1)
ezplot(e)
```

Now publish the report to HTML, by selecting File ► Save and Publish to HMTL (or just File ► Publish to HMTL if you have already saved the M-file), or by clicking the publish button:

The M-file is evaluated and the report is presented in HTML form in a new window. The report is also saved

to a file with the same name as your M-file but with an
html file type. It includes the cell titles (the text after the
double %%), the code itself, the output of the code, and
any figures generated. You can change this default
behavior in the File ▶ Preferences menu, under the
Editor/Debugger: Publishing section.

To run the M-file without publishing the results, simply
click the run button, as usual, or select Cell ▶ Evaluate
Entire Cell. Individual cells can also be evaluated.

Additional descriptive text can be added as plain
comments (one %) after the cell divider but before any
commands. The text can be marked in various styles
(bold, monospaced, TeX equations, and bullet lists, for
example). See the Cell ▶ Insert Text Markup ▶ ...
menu for a complete list.

To add descriptive text without starting a new report
section, start with a cell divider that has no title (a line
containing just %%). This creates a new cell, but it appears
in the same section of the report as the cell before it.

21. Code Development Tools

The Current Directory window provides a pull-down
menu with seven different reports that it can generate.
These tools are described in the seven sections of this
chapter, below.

The Current Directory window has two modes of display,
the classic view and the visual directory view. In the
visual directory view, you can click on a filename in the
Current Directory window to edit it. If cell publishing
has been used to publish the results of an M-file to an

HTML file, a link to the published report will appear next to the filename. A one-line description of each M-file is listed.

21.1 M-lint code check report

Navigate to the directory where you created the ddomloops M-file (see Chapter 8). On Microsoft Windows, this is your work directory by default. In the Current Directory window, select the M-Lint Code Check Report. The report examines all M-files in the directory and checks them for suspicious constructs. Scroll down to the report on ddomloops.m, and note that one warning is listed:

<u>5:</u> The value assigned here to variable 'm' is never used.

Click on the underlined <u>5:</u>. The Editor window opens the ddomloops.m file and highlights line 5:

```
[m n] = size(A) ;
```

The variable m is assigned by this statement, but not used. This is not an error, just a warning. It does remind you that ddomloops is only intended for square matrices, however. This is a good reminder, because no test is made to ensure the matrix is square. Try:

```
ddomloops(ones(2,3))
```

An obscure error occurs because the non-existent entry A(3,3) is referenced. This is not a reliable function.

Save a copy of your original `ddomloops.m` file, and call it `ddomloops_orig.m`. You will need it for the example in Section 21.6.

Add the following code to `ddomloops` just after line 5:

```
if (m ~= n)
    error('A must be square') ;
end
```

Rerun the M-lint report by clicking the Refresh button:

The warning has gone away, and your code is more reliable. Try `ddomloops(ones(2,3))` again. It correctly reports an error that A must be square.

21.2 TODO/FIXME report

The TODO/FIXME Report lists all lines in an M-file containing the words TODO, FIXME, or NOTE, along with the line numbers in which they appear. Clicking the line number brings up the editor at that line. This is useful during incremental development of a large project.

21.3 Help report

The Help Report examines each M-file in the current directory for the comment lines that appear when you type `help` or `doc` followed by the M-file name. Here is its report on `ddomloops`:

```
B = ddomloops(A) returns a diagonally

    B = ddomloops(A) returns a diagonally
    dominant matrix B by modifying the
    diagonal of A.
```

No example
No see-also line
No copyright line

The first line in the report is the description line, which is the first line after the `function` statement itself (if the line is a comment line). The MATLAB convention is for the first comment line to be a stand-alone one-line description of the function, starting with the name of the function in all capital letters. Edit `ddomloops` and add a new description line, as the second line in the file:

```
%DDOMLOOPS make matrix diagonally dominant
```

The Help Report also complains that there is no example, no see-also line, and no copyright line. An example starts with a comment line that starts with the word `example` or `Example` and ends at the next blank comment line. The see-also line is a comment line that starts with the words `See also`. The copyright line is a comment that starts with the word `Copyright`. All of these constructs are optional, of course, but adding them to the M-file makes the code easier to use. After the last comment line, add the following comments:

```
%
% Example
%       A = [1 0 ; 4 1]
%       B = ddomloops(A)
%       B is the same as A, except B(2,2)
%       is slightly greater than 4.
```

```
%
% See also DDOM, DIAGDOM.
```

Finally, add a blank line (not a comment), and then the line:

```
% Copyright 2004, Me.
```

The function names DDOM and DIAGDOM appear in upper case, so that they can be recognized as function names. Rerun the Help Report. You will see all of these constructs listed in the report. Type help ddomloops or doc ddomloops in the Command Window. You should see ddom and diagdom underlined and in blue as active links. Click on them, and you will see the corresponding help or doc for those functions (assuming you created them in Chapters 7 and 8).

21.4 Contents report

The Contents Report generates a special file called Contents.m that summarizes all of the M-files in the current directory. Select it from the menu, and scroll down until you see your modified ddomloops function. Its name is followed by its one-line description, generated automatically from the description line in ddomloops.m. You can edit the Contents.m file to add more description, and then click the refresh button to generate a new Contents Report. Any discrepancies are reported to you. For example, if you edit the one-line description in Contents.m, but not in the corresponding M-file, a warning will appear and you will have the opportunity to fix the discrepancy.

Type the command help *directory* where *directory* is the name of the current directory. This use of the help command prints the Contents.m listing, and highlights the name of each function. Click on ddomloops in the list, and the help ddomloops information will appear. Many of MATLAB's functions are implemented as M-files and are documented in the same way that you have documented your current directory. For example, help general lists the Contents.m file of the directory MATLAB/toolbox/matlab/general (where MATLAB is the directory in which MATLAB is installed).

Create a directory entitled diagonal_dominance and place all of the related M-files and mexFunctions in this directory. Add the diagonal_dominance directory to your path (see Section 7.7). Now, whatever your current directory is help diagonal_dominance will list these files, and the ddom, ddomloops, and diagdom functions will always be available to you.

21.5 Dependency report

For each M-file in the current directory, the Dependency Report lists the M-files and mexFunctions that it relies on, and which M-files rely on it. Create an M-file script in the diagonal_dominance directory called simple.m:

```
A = [1 2 ; 3 0]
B = ddomloops(A)
C = diagdom(A)
```

Run the dependency report. simple is listed as a parent of its child function ddomloops. The mexFunction diagdom is listed as a child of simple. diagdom itself does not appear in the list because it is not an M-file.

21.6 File comparison report

The File Comparison Report is very useful in tracking changes to your code as you develop it. Select this report, and scroll down until you see your original ddomloops_orig file. Click <file 1>. Next, find your new ddomloops and click <file 2>. A color-coded side-by-side display of these two functions is displayed. Plain gray text is code that is identical between the two files. Pink highlighting denotes lines that differ between the two files. Green highlighting denotes lines that appear in one file but not the other.

21.7 Profile and coverage report

MATLAB provides an M-file profiler that lets you see how much computation time each line of an M-file uses. Select Desktop ▸ Profiler or type profile viewer. Try this example. Type in a M-file script, ddomtest.m:

```
A = rand(1000) ;
B = ddomloops(A) ;
```

Type ddomtest in the text window entitled Run this code and hit enter. A short table appears with the number of calls and time spent in each function. Most of the time is spent in ddomloops. Click on the function name and you are provided a lengthy description of the time spent in ddomloops. This report is useful for improving code performance and for debugging. Untested lines of code could harbor a bug.

The Coverage Report provides a short overview of the profile coverage of each file in a directory. Selecting it shows that ddomtest was fully exercised (100% coverage), but a few lines of code in ddomloops were not tested. The code you added to check for rectangular matrices was not tested, and the case when the diagonal entry $A(i,i)$ is negative was not tested.

22. Help Topics

There are many MATLAB functions and features that cannot be included in this Primer. Listed in the following tables are some of the MATLAB functions and operators, grouped by subject area. You can browse through these lists and use the online help facility, or consult the online documents for more detailed information on the functions, operators, and special characters. Open the Help Browser to `Help: MATLAB: Functions -- Categorical List`.

The `help` command lists help information in the MATLAB Command window. The tables are derived from the MATLAB 7 (R14) `help` command. Typing `help` alone will provide a listing of the major MATLAB directories, similar to the following table. Typing `help topic`, where *topic* is an entry in the left column of the table, will display a description of the topic. For example, `help general` will display on your Command window a plain text version of Section 22.1. Typing `help ops` will display Section 22.2, starting on page 144, and so on.

The `doc` command opens the MATLAB help browser. It display the M-file help, just as the help command, if the command has no HTML reference page. Try `doc general` or `doc ops`.

Each topic is discussed in a single subsection. The page number for each subsection is also listed in the following table.

Help topics		page
`general`	General purpose commands	142
`ops`	Operators and special characters	144
`lang`	Programming language constructs	147
`elmat`	Elementary matrices and matrix manipulation	149
`elfun`	Elementary math functions	151
`specfun`	Specialized math functions	153
`matfun`	Matrix functions - linear algebra	155
`datafun`	Data analysis & Fourier transforms	157
`polyfun`	Interpolation and polynomials	158
`funfun`	Function functions & ODE solvers	160
`sparfun`	Sparse matrices	162
`scribe`	Annotation and plot editing	164
`graph2d`	Two-dimensional graphs	164
`graph3d`	Three-dimensional graphs	165
`specgraph`	Specialized graphs	168
`graphics`	Handle Graphics	171
`uitools`	Graphical user interface tools	173
`strfun`	Character strings	176
`imagesci`	Image, scientific data input/output	178
`iofun`	File input/output	179
`audiovideo`	Audio and video support	182
`timefun`	Time and dates	183
`datatypes`	Data types and structures	183
`verctrl`	Version control	187
`codetools`	Creating and debugging code	187
`helptools`	Help commands	188
`winfun`	Microsoft Windows functions	189
`demos`	Examples and demonstrations	190
`local`	Preferences	190
`symbolic`	Symbolic Math Toolbox	191

22.1 General purpose commands

`help general`

General information

`syntax`	Help on MATLAB command syntax
`demo`	Run demonstrations
`ver`	MATLAB, Simulink, & toolbox version
`version`	MATLAB version information

Managing the workspace

`who`	List current variables
`whos`	List current variables, long form
`clear`	Clear variables, functions from memory
`pack`	Consolidate workspace memory
`load`	Load variables from MAT- or ASCII file
`save`	Save variables to MAT- or ASCII file
`saveas`	Save figure or model to file
`memory`	Help for memory limitations
`recycle`	Recycle folder option for deleted files
`quit`	Quit MATLAB session
`exit`	Exit from MATLAB

Managing commands and functions

`what`	List MATLAB-specific files in directory
`type`	List M-file
`open`	Open files by extension
`which`	Locate functions and files
`pcode`	Create pre-parsed P-file
`mex`	Compile MEX-function
`inmem`	List functions in memory
`namelengthmax`	Max length of function or variable name

Managing the search path

path	Get/set search path
addpath	Add directory to search path
rmpath	Remove directory from search path
rehash	Refresh function and file system caches
import	Import Java packages into current scope
finfo	Identify file type
genpath	Generate recursive toolbox path
savepath	Save MATLAB path in pathdef.m file

Managing the Java search path

javaaddpath	Add directories to the dynamic Java path
javaclasspath	Get and set Java path
javarmpath	Remove dynamic Java path directory

Controlling the Command window

echo	Echo commands in M-files
more	Paged output in command window
diary	Save text of MATLAB session
format	Set output format
beep	Produce beep sound
desktop	Start and query the MATLAB Desktop
preferences	MATLAB user preferences dialog

Debugging

debug	List debugging commands

Locate dependent functions of an M-file

depfun	Find dependent functions of M- or P-file
depdir	Find dependent directories of M or P-file

Operating system commands

cd	Change current working directory
copyfile	Copy file or directory
movefile	Move file or directory
delete	Delete file or graphics object
pwd	Show (print) current working directory
dir	List directory
ls	List directory
fileattrib	Set/get attributes of files and directories
isdir	True if argument is a directory
mkdir	Make new directory
rmdir	Remove directory
getenv	Get environment variable
!	Execute operating system command
dos	Execute DOS command and return result
unix	Execute Unix command and return result
system	Execute system command, return result
perl	Execute Perl command and return result
computer	Computer type
isunix	True for Unix version of MATLAB
ispc	True for Windows version of MATLAB

Loading and calling shared libraries

calllib	Call a function in an external library
libpointer	Create pointer for external libraries
libstruct	Create structure ptr. for external libraries
libisloaded	True if specified shared library is loaded
loadlibrary	Load a shared library into MATLAB
libfunctions	Info. on functions in external library
libfunctionsview	View functions in external library
unloadlibrary	Unload a shared library
java	Using Java from within MATLAB
usejava	True if Java feature supported

22.2 Operators and special characters

`help ops`

Arithmetic operators (help arith, help slash)		
plus	Plus	+
uplus	Unary plus	+
minus	Minus	−
uminus	Unary minus	−
mtimes	Matrix multiply	*
times	Array multiply	.*
mpower	Matrix power	^
power	Array power	.^
mldivide	Backslash or left matrix divide	\
mrdivide	Slash or right matrix divide	/
ldivide	Left array divide	.\
rdivide	Right array divide	./
kron	Kronecker tensor product	kron

Relational operators (help relop)		
eq	Equal	==
ne	Not equal	~=
lt	Less than	<
gt	Greater than	>
le	Less than or equal	<=
ge	Greater than or equal	>=

Logical operators		
	Short-circuit logical AND	&&
	Short-circuit logical OR	||
and	Logical AND	&
or	Logical OR	|
not	Logical NOT	~
xor	Logical EXCLUSIVE OR	
any	True if any element of vector is nonzero	
all	True if all elements of vector are nonzero	

Special characters

`colon`	Colon	:
`paren`	Parentheses and subscripting	( )
`paren`	Brackets	[ ]
`paren`	Braces and subscripting	{ }
`punct`	Function handle creation	@
`punct`	Decimal point	.
`punct`	Structure field access	.
`punct`	Parent directory	..
`punct`	Continuation	...
`punct`	Separator	,
`punct`	Semicolon	;
`punct`	Comment	%
`punct`	Operating system command	!
`punct`	Assignment	=
`punct`	Quote	'
`transpose`	Transpose	.'
`ctranspose`	Complex conjugate transpose	'
`horzcat`	Horizontal concatenation	[,]
`vertcat`	Vertical concatenation	[;]
`subsasgn`	Subscripted assignment	( ) { }
`subsref`	Subscripted reference	( ) { }
`subsindex`	Subscript index	

Bitwise operators

`bitand`	Bit-wise AND
`bitcmp`	Complement bits
`bitor`	Bit-wise OR
`bitmax`	Maximum floating-point integer
`bitxor`	Bit-wise EXCLUSIVE OR
`bitset`	Set bit
`bitget`	Get bit
`bitshift`	Bit-wise shift

Set operators	
union	Set union
unique	Set unique
intersect	Set intersection
setdiff	Set difference
setxor	Set exclusive-or
ismember	True for set member

22.3 Programming language constructs
help lang

Control flow	
if	Conditionally execute statements
else	When if condition fails
elseif	When if failed and condition is true
end	Scope of for, while, switch, try, if
for	Repeat specific number of times
while	Repeat an indefinite number of times
break	Terminate while or for loop
continue	Pass control to next iteration of a loop
switch	Switch among several cases
case	switch statement case
otherwise	Default switch statement case
try	Begin try block
catch	Begin catch block
return	Return to invoking function
error	Display mesage and abort function
rethrow	Reissue error

Evaluation and execution	
eval	Execute MATLAB expression in string
evalc	eval with capture
feval	Execute function specified by string
evalin	Evaluate expression in workspace
builtin	Execute built-in function
assignin	Assign variable in workspace
run	Run script

Scripts, functions, and variables

script	About MATLAB scripts and M-files
function	Add new function
global	Define global variable
persistent	Define persistent variable
mfilename	Name of currently executing M-file
lists	Comma separated lists
exist	Check if variables or functions defined
isglobal	True for global variables *(obsolete)*
mlock	Prevent M-file from being cleared
munlock	Allow M-file to be cleared
mislocked	True if M-file cannot be cleared
precedence	Operator precedence in MATLAB
isvarname	Check for a valid variable name
iskeyword	Check if input is a keyword
javachk	Validate level of Java support
genvarname	MATLAB variable name from string

Argument handling

nargchk	Validate number of input arguments
nargoutchk	Validate number of output arguments
nargin	Number of function input arguments
nargout	Number of function output arguments
varargin	Variable length input argument list
varargout	Variable length output argument list
inputname	Input argument name

Message display and interactive input

warning	Display warning message
lasterr	Last error message
lastwarn	Last warning message
disp	Display array
display	Display array
intwarning	Controls state of the 4 integer warnings
input	Prompt for user input
keyboard	Invoke keyboard from M-file

22.4 Elementary matrices and matrix manipulation

help elmat

Elementary matrices	
zeros	Zeros array
ones	Ones array
eye	Identity matrix
repmat	Replicate and tile array
rand	Uniformly distributed random numbers
randn	Normally distributed random numbers
linspace	Linearly spaced vector
logspace	Logarithmically spaced vector
freqspace	Frequency spacing for frequency response
meshgrid	x and y arrays for 3-D plots
accumarray	Construct an array with accumulation
:	Regularly spaced vector and array index

Basic array information	
size	Size of matrix
length	Length of vector
ndims	Number of dimensions
numel	Number of elements
disp	Display matrix or text
isempty	True for empty matrix
isequal	True if arrays are numerically equal
isequalwithequalnans	True if arrays are numerically equal (assuming nan==nan)

Array utility functions	
isscalar	True for scalar
isvector	True for vector

Matrix manipulation

cat	Concatenate arrays
reshape	Change size
diag	Diagonal matrices; diagonals of matrix
blkdiag	Block diagonal concatenation
tril	Extract lower triangular part
triu	Extract upper triangular part
fliplr	Flip matrix in left/right direction
flipud	Flip matrix in up/down direction
flipdim	Flip matrix along specified dimension
rot90	Rotate matrix 90 degrees
:	Regularly spaced vector and array index
find	Find indices of nonzero elements
end	Last index
sub2ind	Linear index from multiple subscripts
ind2sub	Multiple subscripts from linear index
ndgrid	Arrays of N-D functions & interpolation
permute	Permute array dimensions
ipermute	Inverse permute array dimensions
shiftdim	Shift dimensions
circshift	Shift array circularly
squeeze	Remove singleton dimensions

Special variables and constants

ans	Most recent answer
eps	Floating-point relative accuracy
realmax	Largest positive floating-point number
realmin	Smallest positive floating-point number
pi	3.1415926535897...
i, j	Imaginary unit
inf	Infinity
nan	Not-a-Number
isnan	True for Not-a-Number
isinf	True for infinite elements
isfinite	True for finite elements

Specialized matrices	
compan	Companion matrix
gallery	Higham test matrices
hadamard	Hadamard matrix
hankel	Hankel matrix
hilb	Hilbert matrix
invhilb	Inverse Hilbert matrix
magic	Magic square
pascal	Pascal matrix
rosser	Symmetric eigenvalue test problem
toeplitz	Toeplitz matrix
vander	Vandermonde matrix
wilkinson	Wilkinson's eigenvalue test matrix

22.5 Elementary math functions

help elfun

Trigonometric (also continued on next page)	
sin	Sine
sind	Sine of argument in degrees
sinh	Hyperbolic sine
asin	Inverse sine
asind	Inverse sine, result in degrees
asinh	Inverse hyperbolic sine
cos	Cosine
cosd	Cosine of argument in degrees
cosh	Hyperbolic cosine
acos	Inverse cosine
acosd	Inverse cosine, result in degrees
acosh	Inverse hyperbolic cosine
tan	Tangent
tand	Tangent of argument in degrees
tanh	Hyperbolic tangent
atan	Inverse tangent
atand	Inverse tangent, result in degrees
atan2	Four quadrant inverse tangent

Trigonometric (continued)

`atanh`	Inverse hyperbolic tangent
`sec`	Secant
`secd`	Secant of argument in degrees
`sech`	Hyperbolic secant
`asec`	Inverse secant
`asecd`	Inverse secant, result in degrees
`asech`	Inverse hyperbolic secant
`csc`	Cosecant
`cscd`	Cosecant of argument in degrees
`csch`	Hyperbolic cosecant
`acsc`	Inverse cosecant
`acscd`	Inverse cosecant, result in degrees
`acsch`	Inverse hyperbolic cosecant
`cot`	Cotangent
`cotd`	Cotangent of argument in degrees
`coth`	Hyperbolic cotangent
`acot`	Inverse cotangent
`acotd`	Inverse cotangent, result in degrees
`acoth`	Inverse hyperbolic cotangent

Exponential

`exp`	Exponential
`expm1`	Compute `exp(x)-1` accurately
`log`	Natural logarithm
`log1p`	Compute `log(1+x)` accurately
`log10`	Common (base 10) logarithm
`log2`	Base 2 logarithm, dissect floating-point
`pow2`	Base 2 power, scale floating-point
`realpow`	Array power with real result (or error)
`reallog`	Natural logarithm of real number
`realsqrt`	Square root of number ≥ 0
`sqrt`	Square root
`nthroot`	Real n^{th} root of real numbers
`nextpow2`	Next higher power of 2

Complex	
abs	Absolute value
angle	Phase angle
complex	Complex from real and imaginary parts
conj	Complex conjugate
imag	Complex imaginary part
real	Complex real part
unwrap	Unwrap phase angle
isreal	True for real array
cplxpair	Sort into complex conjugate pairs

Rounding and remainder	
fix	Round towards zero
floor	Round towards minus infinity
ceil	Round towards plus infinity
round	Round towards nearest integer
mod	Modulus (remainder after division)
rem	Remainder after division
sign	Signum

22.6 Specialized math functions
help specfun

Number theoretic functions	
factor	Prime factors
isprime	True for prime numbers
primes	Generate list of prime numbers
gcd	Greatest common divisor
lcm	Least common multiple
rat	Rational approximation
rats	Rational output
perms	All possible permutations
nchoosek	All combinations of N choose K
factorial	Factorial function

Specialized math functions

airy	Airy functions
besselj	Bessel function of the first kind
bessely	Bessel function of the second kind
besselh	Bessel function of 3rd kind (Hankel function)
besseli	Modified Bessel function of the 1st kind
besselk	Modified Bessel function of the 2nd kind
beta	Beta function
betainc	Incomplete beta function
betaln	Logarithm of beta function
ellipj	Jacobi elliptic functions
ellipke	Complete elliptic integral
erf	Error function
erfc	Complementary error function
erfcx	Scaled complementary error function
erfinv	Inverse error function
expint	Exponential integral function
gamma	Gamma function
gammainc	Incomplete gamma function
gammaln	Logarithm of gamma function
psi	Psi (polygamma) function
legendre	Associated Legendre function
cross	Vector cross product
dot	Vector dot product

Coordinate transforms

cart2sph	Cartesian to spherical coordinates
cart2pol	Cartesian to polar coordinates
pol2cart	Polar to Cartesian coordinates
sph2cart	Spherical to Cartesian coordinates
hsv2rgb	Convert HSV colors to RGB
rgb2hsv	Convert RGB colors to HSV

22.7 Matrix functions — numerical linear algebra

`help matfun`

Matrix analysis	
`norm`	Matrix or vector norm
`normest`	Estimate the matrix 2-norm
`rank`	Matrix rank
`det`	Determinant
`trace`	Sum of diagonal elements
`null`	Null space
`orth`	Orthogonalization
`rref`	Reduced row echelon form
`subspace`	Angle between two subspaces

Linear equations	
`\ and /`	Linear equation solution (`help slash`)
`linsolve`	Linear equation solution, extra control
`inv`	Matrix inverse
`rcond`	LAPACK reciprocal condition estimator
`cond`	Condition number
`condest`	1-norm condition number estimate
`normest1`	1-norm estimate
`chol`	Cholesky factorization
`cholinc`	Incomplete Cholesky factorization
`lu`	LU factorization
`luinc`	Incomplete LU factorization
`qr`	Orthogonal-triangular decomposition
`lsqnonneg`	Linear least squares (≥ 0 constraints)
`pinv`	Pseudoinverse
`lscov`	Least squares with known covariance

Eigenvalues and singular values

eig	Eigenvalues and eigenvectors
svd	Singular value decomposition
gsvd	Generalized singular value decomp.
eigs	A few eigenvalues
svds	A few singular values
poly	Characteristic polynomial
polyeig	Polynomial eigenvalue problem
condeig	Condition number of eigenvalues
hess	Hessenberg form
qz	QZ factoriz. for generalized eigenvalues
ordqz	Reordering of eigenvalues in QZ
schur	Schur decomposition
ordschur	Sort eigenvalues in Schur decomposition

Matrix functions

expm	Matrix exponential
logm	Matrix logarithm
sqrtm	Matrix square root
funm	Evaluate general matrix function

Factorization utilities

qrdelete	Delete column from QR factorization
qrinsert	Insert column in QR factorization
rsf2csf	Real block diagonal to complex diagonal
cdf2rdf	Complex diagonal to real block diagonal
balance	Diagonal scaling for eigenvalue accuracy
planerot	Givens plane rotation
cholupdate	rank 1 update to Cholesky factorization
qrupdate	rank 1 update to QR factorization

22.8 Data analysis, Fourier transforms
`help datafun`

Basic operations	
max	Largest component
min	Smallest component
mean	Average or mean value
median	Median value
std	Standard deviation
var	Variance
sort	Sort in ascending order
sortrows	Sort rows in ascending order
sum	Sum of elements
prod	Product of elements
hist	Histogram
histc	Histogram count
trapz	Trapezoidal numerical integration
cumsum	Cumulative sum of elements
cumprod	Cumulative product of elements
cumtrapz	Cumulative trapezoidal num. integration

Finite differences	
diff	Difference and approximate derivative
gradient	Approximate gradient
del2	Discrete Laplacian

Correlation	
corrcoef	Correlation coefficients
cov	Covariance matrix
subspace	Angle between subspaces

Filtering and convolution	
filter	One-dimensional digital filter
filter2	Two-dimensional digital filter
conv	Convolution, polynomial multiplication
conv2	Two-dimensional convolution
convn	N-dimensional convolution
deconv	Deconvolution and polynomial division
detrend	Linear trend removal

Fourier transforms	
fft	Discrete Fourier transform
fft2	2-D discrete Fourier transform
fftn	N-D discrete Fourier transform
ifft	Inverse discrete Fourier transform
ifft2	2-D inverse discrete Fourier transform
ifftn	N-D inverse discrete Fourier transform
fftshift	Shift zero-freq. component to center
ifftshift	Inverse fftshift

22.9 Interpolation and polynomials
help polyfun

Data interpolation	
pchip	Piecewise cubic Hermite interpol. poly.
interp1	1-D interpolation (table lookup)
interp1q	Quick 1-D linear interpolation
interpft	1-D interpolation using FFT method
interp2	2-D interpolation (table lookup)
interp3	3-D interpolation (table lookup)
interpn	N-D interpolation (table lookup)
griddata	2-D data gridding and surface fitting
griddata3	3-D data gridding & hypersurface fitting
griddatan	N-D data gridding &hypersurface fitting

Spline interpolation

spline	Cubic spline interpolation
ppval	Evaluate piecewise polynomial

Geometric analysis

delaunay	Delaunay triangulation
delaunay3	3-D Delaunay tessellation
delaunayn	N-D Delaunay tessellation
dsearch	Search Delaunay triangulation
dsearchn	Search N-D Delaunay tessellation
tsearch	Closest triangle search
tsearchn	N-D closest triangle search
convhull	Convex hull
convhulln	N-D convex hull
voronoi	Voronoi diagram
voronoin	N-D Voronoi diagram
inpolygon	True for points inside polygonal region
rectint	Rectangle intersection area
polyarea	Area of polygon

Polynomials

roots	Find polynomial roots
poly	Convert roots to polynomial
polyval	Evaluate polynomial
polyvalm	Evaluate polynomial (matrix argument)
residue	Partial-fraction expansion (residues)
polyfit	Fit polynomial to data
polyder	Differentiate polynomial
polyint	Integrate polynomial analytically
conv	Multiply polynomials
deconv	Divide polynomials

22.10 Function functions and ODEs

`help funfun`

Optimization and root finding	
`fminbnd`	Scalar bounded nonlinear minimization
`fminsearch`	Multidimensional unconstrained nonlinear minimization (Nelder-Mead)
`fzero`	Scalar nonlinear zero finding

Optimization option handling	
`optimset`	Set optimization `options` structure
`optimget`	Get optimization parameters

Numerical integration (quadrature)	
`quad`	Numerical integration, low order method
`quadl`	Numerical integration, high order method
`dblquad`	Numerically evaluate double integral
`triplequad`	Numerically evaluate triple integral
`quadv`	Vectorized `quad`

Plotting	
`ezplot`	Easy-to-use function plotter
`ezplot3`	Easy-to-use 3-D parametric curve plotter
`ezpolar`	Easy-to-use polar coordinate plotter
`ezcontour`	Easy-to-use contour plotter
`ezcontourf`	Easy-to-use filled contour plotter
`ezmesh`	Easy-to-use 3-D mesh plotter
`ezmeshc`	Easy-to-use mesh/contour plotter
`ezsurf`	Easy-to-use 3-D colored surface plotter
`ezsurfc`	Easy-to-use surf/contour plotter
`fplot`	Plot function

Inline function object	
`inline`	Construct `inline` function object
`argnames`	Argument names
`formula`	Function formula
`char`	Convert `inline` object to `char` array

Initial value problem solvers for ODEs

ode45	Solve non-stiff differential equations, medium order method (Try this first)
ode23	Solve non-stiff ODEs low order method
ode113	Solve non-stiff ODEs, variable order
ode23t	Solve moderately stiff ODEs and DAEs Index 1, trapezoidal rule
ode15s	Solve stiff ODEs and DAEs Index 1, variable order method
ode23s	Solve stiff ODEs, low order method
ode23tb	Solve stiff ODEs, low order method

Initial value problem, fully implicit ODEs/DAEs

decic	Compute consistent initial conditions
ode15i	Solve implicit ODEs or DAEs Index 1

Initial value problem solvers for DDEs

dde23	Solve delay differential equations

Boundary value problem solver for ODEs

bvp4c	Solve two-point boundary value ODEs

1-D Partial differential equation solver

pdepe	Solve initial-boundary value PDEs

ODE, DDE, BVP option handling

odeset	Create/alter ODE options structure
odeget	Get ODE options parameters
ddeset	Create/alter DDE options structure
ddeget	Get DDE options parameters
bvpset	Create/alter BVP options structure
bvpget	Get BVP options parameters

ODE, DAE, DDE, PDE input & output functions	
deval	Evaluate solution of differential equation
odextend	Extend solutions of differential equation
odeplot	Time series ODE output function
odephas2	2-D phase plane ODE output function
odephas3	3-D phase plane ODE output function
odeprint	ODE output function
bvpinit	Forms the initial guess for bvp4c
pdeval	Evaluates solution computed by pdepe

22.11 Sparse matrices

help sparfun

Elementary sparse matrices	
speye	Sparse identity matrix
sprand	Uniformly distributed random matrix
sprandn	Normally distributed random matrix
sprandsym	Sparse random symmetric matrix
spdiags	Sparse matrix formed from diagonals

Full to sparse conversion	
sparse	Create sparse matrix
full	Convert sparse matrix to full matrix
find	Find indices of nonzero elements
spconvert	Create sparse matrix from triplets

Working with sparse matrices	
nnz	Number of nonzero matrix elements
nonzeros	Nonzero matrix elements
nzmax	Space allocated for nonzero elements
spones	Replace nonzero elements with ones
spalloc	Allocate space for sparse matrix
issparse	True for sparse matrix
spfun	Apply function to nonzero elements
spy	Visualize sparsity pattern

Reordering algorithms

`colamd`	Column approximate minimum degree
`symamd`	Approximate minimum degree
`symrcm`	Symmetric reverse Cuthill-McKee
`colperm`	Column permutation
`randperm`	Random permutation
`dmperm`	Dulmage-Mendelsohn permutation
`lu`	UMFPACK reordering (with 4 outputs)

Linear algebra

`eigs`	A few eigenvalues, using ARPACK
`svds`	A few singular values, using `eigs`
`luinc`	Incomplete LU factorization
`cholinc`	Incomplete Cholesky factorization
`normest`	Estimate the matrix 2-norm
`condest`	1-norm condition number estimate
`sprank`	Structural rank

Linear equations (iterative methods)

`pcg`	Preconditioned conjugate gradients
`bicg`	Biconjugate gradients method
`bicgstab`	Biconjugate gradients stabilized method
`cgs`	Conjugate gradients squared method
`gmres`	Generalized minimum residual method
`lsqr`	Conjugate gradients on normal equations
`minres`	Minimum residual method
`qmr`	Quasi-minimal residual method
`symmlq`	Symmetric LQ method

Operations on graphs (trees)

`treelayout`	Lay out tree or forest
`treeplot`	Plot picture of tree
`etree`	Elimination tree
`etreeplot`	Plot elimination tree
`gplot`	Plot graph, as in "graph theory"

Miscellaneous	
symbfact	Symbolic factorization analysis
spparms	Set parameters for sparse matrix routines
spaugment	Form least squares augmented system

22.12 Annotation and plot editting

help scribe

Graph annotation	
annotation	Create annotation objects for figures
colorbar	Display colour bar (color scale)
legend	Graph legend

22.13 Two-dimensional graphs

help graph2d

Elementary x-y graphs	
plot	Linear plot
loglog	Log-log scale plot
semilogx	Semi-log scale plot
semilogy	Semi-log scale plot
polar	Polar coordinate plot
plotyy	Graphs with y tick labels on left & right

Axis control	
axis	Control axis scaling and appearance
zoom	Zoom in and out on a 2-D plot
grid	Grid lines
box	Axis box
hold	Hold current graph
axes	Create axes in arbitrary positions
subplot	Create axes in tiled positions

Hard copy and printing	
print	Print graph, Simulink sys.; save to M-file
printopt	Printer defaults
orient	Set paper orientation

Graph annotation	
`plotedit`	Tools for editing and annotating plots
`title`	Graph title
`xlabel`	*x*-axis label
`ylabel`	*y*-axis label
`texlabel`	TeX format from string
`text`	Text annotation
`gtext`	Place text with mouse

22.14 Three-dimensional graphs
`help graph3d`

Elementary 3-D plots	
`plot3`	Plot lines and points in 3-D space
`mesh`	3-D mesh surface
`surf`	3-D colored surface
`fill3`	Filled 3-D polygons

Color control	
`colormap`	Color look-up table
`caxis`	Pseudocolor axis scaling
`shading`	Color shading mode
`hidden`	Mesh hidden line removal mode
`brighten`	Brighten or darken color map
`colordef`	Set color defaults
`graymon`	Graphics defaults for grayscale monitors

Lighting	
`surfl`	3-D shaded surface with lighting
`lighting`	Lighting mode
`material`	Material reflectance mode
`specular`	Specular reflectance
`diffuse`	Diffuse reflectance
`surfnorm`	Surface normals

Color maps

hsv	Hue-saturation-value color map
hot	Black-red-yellow-white color map
gray	Linear grayscale color map
bone	Grayscale with tinge of blue color map
copper	Linear copper-tone color map
pink	Pastel shades of pink color map
white	All-white color map
flag	Alternating red, white, blue, and black
lines	Color map with the line colors
colorcube	Enhanced color-cube color map
vga	Windows colormap for 16 colors
jet	Variant of HSV
prism	Prism color map
cool	Shades of cyan and magenta color map
autumn	Shades of red and yellow color map
spring	Shades of magenta and yellow color map
winter	Shades of blue and green color map
summer	Shades of green and yellow color map

Axis control

axis	Control axis scaling and appearance
zoom	Zoom in and out on a 2-D plot
grid	Grid lines
box	Axis box
hold	Hold current graph
axes	Create axes in arbitrary positions
subplot	Create axes in tiled positions
daspect	Data aspect ratio
pbaspect	Plot box aspect ratio
xlim	x limits
ylim	y limits
zlim	z limits

Transparency

alpha	Transparency (alpha) mode
alphamap	Transparency (alpha) look-up table
alim	Transparency (alpha) scaling

Viewpoint control

view	3-D graph viewpoint specification
viewmtx	View transformation matrix
rotate3d	Interactively rotate view of 3-D plot

Camera control

campos	Camera position
camtarget	Camera target
camva	Camera view angle
camup	Camera up vector
camproj	Camera projection

High-level camera control

camorbit	Orbit camera
campan	Pan camera
camdolly	Dolly camera
camzoom	Zoom camera
camroll	Roll camera
camlookat	Move camera and target to view objects
cameratoolbar	Interactively manipulate camera

High-level light control

camlight	Creates or sets position of a light
lightangle	Spherical position of a light

Hard copy and printing

print	Print graph, Simulink sys.; save to M-file
printopt	Printer defaults
orient	Set paper orientation
vrml	Save graphics to VRML 2.0 file

Graph annotation	
title	Graph title
xlabel	*x*-axis label
ylabel	*y*-axis label
zlabel	*z*-axis label
text	Text annotation
gtext	Mouse placement of text
plotedit	Graph editing and annotation tools

22.15 Specialized graphs

help specgraph

Specialized 2-D graphs	
area	Filled area plot
bar	Bar graph
barh	Horizontal bar graph
comet	Comet-like trajectory
compass	Compass plot
errorbar	Error bar plot
ezplot	Easy-to-use function plotter
ezpolar	Easy-to-use polar coordinate plotter
feather	Feather plot
fill	Filled 2-D polygons
fplot	Plot function
hist	Histogram
pareto	Pareto chart
pie	Pie chart
plotmatrix	Scatter plot matrix
rose	Angle histogram plot
scatter	Scatter plot
stem	Discrete sequence or "stem" plot
stairs	Stairstep plot

Contour and 2½-D graphs

contour	Contour plot
contourf	Filled contour plot
contour3	3-D contour plot
clabel	Contour plot elevation labels
ezcontour	Easy-to-use contour plotter
ezcontourf	Easy-to-use filled contour plotter
pcolor	Pseudocolor (checkerboard) plot
voronoi	Voronoi diagram

Specialized 3-D graphs

bar3	3-D bar graph
bar3h	Horizontal 3-D bar graph
comet3	3-D comet-like trajectories
ezgraph3	General-purpose surface plotter
ezmesh	Easy-to-use 3-D mesh plotter
ezmeshc	Easy-to-use mesh/contour plotter
ezplot3	Easy-to-use 3-D parametric curve plotter
ezsurf	Easy-to-use 3-D colored surface plotter
ezsurfc	Easy-to-use surf/contour plotter
meshc	Combination mesh/contour plot
meshz	3-D mesh with curtain
pie3	3-D pie chart
ribbon	Draw 2-D lines as ribbons in 3-D
scatter3	3-D scatter plot
stem3	3-D stem plot
surfc	Combination surf/contour plot
trisurf	Triangular surface plot
trimesh	Triangular mesh plot
waterfall	Waterfall plot

Color-related functions

spinmap	Spin color map
rgbplot	Plot color map
colstyle	Parse color and style from string
ind2rgb	Convert indexed image to RGB image

Volume and vector visualization

vissuite	Visualization suite
isosurface	Isosurface extractor
isonormals	Isosurface normals
isocaps	Isosurface end caps
isocolors	Isosurface and patch colors
contourslice	Contours in slice planes
slice	Volumetric slice plot
streamline	Streamlines from 2-D or 3-D vector data
stream3	3-D streamlines
stream2	2-D streamlines
quiver3	3-D quiver plot
quiver	2-D quiver plot
divergence	Divergence of a vector field
curl	Curl and angular velocity of vector field
coneplot	3-D cone plot
streamtube	3-D stream tube
streamribbon	3-D stream ribbon
streamslice	Streamlines in slice planes
streamparticles	Display stream particles
interpstreamspeed	Interpolate streamlines from speed
subvolume	Extract subset of volume dataset
reducevolume	Reduce volume dataset
volumebounds	Returns x,y,z, & color limits for volume
smooth3	Smooth 3-D data
reducepatch	Reduce number of patch faces
shrinkfaces	Reduce size of patch faces

Movies and animation

moviein	Initialize movie frame memory
getframe	Get movie frame
movie	Play recorded movie frames
rotate	Rotate about specified orgin & direction
frame2im	Convert movie frame to indexed image
im2frame	Convert index image into movie format

Image display and file I/O	
image	Display image
imagesc	Scale data and display as image
colormap	Color look-up table
gray	Linear grayscale color map
contrast	Grayscale color map to enhance contrast
brighten	Brighten or darken color map
colorbar	Display color bar (color scale)
imread	Read image from graphics file
imwrite	Write image to graphics file
imfinfo	Information about graphics file
im2java	Convert image to Java image

Solid modeling	
cylinder	Generate cylinder
sphere	Generate sphere
ellipsoid	Generate ellipsoid
patch	Create patch
surf2patch	Convert surface data to patch data

22.16 Handle Graphics

help graphics

Figure window creation and control	
figure	Create figure window
gcf	Get handle to current figure
clf	Clear current figure
shg	Show graph window
close	Close figure
refresh	Refresh figure
refreshdata	Refresh data plotted in figure
openfig	Open new or raise copy of saved figure

Axis creation and control

`subplot`	Create axes in tiled positions
`axes`	Create axes in arbitrary positions
`gca`	Get handle to current axes
`cla`	Clear current axes
`axis`	Control axis scaling and appearance
`box`	Axis box
`caxis`	Control pseudocolor axis scaling
`hold`	Hold current graph
`ishold`	Return hold state

Handle Graphics objects

`figure`	Create figure window
`axes`	Create axes
`line`	Create line
`text`	Create text
`patch`	Create patch
`rectangle`	Create rectangle or ellipse
`surface`	Create surface
`image`	Create image
`light`	Create light
`uicontrol`	Create user interface control
`uimenu`	Create user interface menu
`uicontextmenu`	Create user interface context menu

Hard copy and printing

`print`	Print graph, Simulink sys.; save to M-file
`printopt`	Printer defaults
`orient`	Set paper orientation

Utilities

`closereq`	Figure close request function
`newplot`	M-file preamble for NextPlot property
`ishandle`	True for graphics handles

Handle Graphics operations	
set	Set object properties
get	Get object properties
reset	Reset object properties
delete	Delete object
gco	Get handle to current object
gcbo	Get handle to current callback object
gcbf	Get handle to current callback figure
drawnow	Flush pending graphics events
findobj	Find objects w/ specified property values
copyobj	Copy graphics object and its children
isappdata	Check if application-defined data exists
getappdata	Get value of application-defined data
setappdata	Set application-defined data
rmappdata	Remove application-defined data

22.17 Graphical user interface tools
help uitools

GUI functions	
uicontrol	Create user interface control
uimenu	Create user interface menu
ginput	Graphical input from mouse
dragrect	Drag XOR rectangles with mouse
rbbox	Rubberband box
selectmoveresize	Select, move, resize, copy objects
waitforbuttonpress	Wait for key/buttonpress
waitfor	Block execution and wait for event
uiwait	Block execution and wait for resume
uiresume	Resume execution of blocked M-file
uistack	Control stacking order of objects
uisuspend	Suspend the interactive state of a figure
uirestore	Restore the interactive state of a figure

GUI design tools

guide	Design GUI
inspect	Inspect object properties
align	Align uicontrols and axes
propedit	Edit property

Dialog boxes

axlimdlg	Axes limits dialog box
dialog	Create dialog figure
errordlg	Error dialog box
helpdlg	Help dialog box
imageview	Show image preview in a figure window
inputdlg	Input dialog box
listdlg	List selection dialog box
menu	Generate menu of choices for user input
movieview	Show movie in figure with replay button
msgbox	Message box
pagedlg	Page position dialog box
pagesetupdlg	Page setup dialog
printdlg	Print dialog box
printpreview	Display preview of figure to be printed
questdlg	Question dialog box
soundview	Show sound in figure and play
uigetpref	Question dialog box with preference
uigetfile	Standard open file dialog box
uiputfile	Standard save file dialog box
uigetdir	Standard open directory dialog box
uisetcolor	Color selection dialog box
uisetfont	Font selection dialog box
uiopen	Show open file dialog and call open
uisave	Show open file dialog and call save
uiload	Show open file dialog and call load
waitbar	Display wait bar
warndlg	Warning dialog box

Menu utilities

makemenu	Create menu structure
menubar	Default setting for MenuBar property
umtoggle	Toggle checked status of uimenu object
winmenu	Create submenu for Window menu item

Toolbar button group utilities

btngroup	Create toolbar button group
btnresize	Resize button group
btnstate	Query state of toolbar button group
btnpress	Button press manager
btndown	Depress button in toolbar button group
btnup	Raise button in toolbar button group

Miscellaneous utilities

allchild	Get all object children
clipboard	Copy/paste to/from system clipboard
edtext	Interactive editing of axes text objects
findall	Find all objects
findfigs	Find figures positioned off screen
getptr	Get figure pointer
getstatus	Get status text string in figure
hidegui	Hide/unhide GUI
listfonts	Get list of available system fonts
movegui	Move GUI to specified part of screen
guihandles	Return a structure of handles
guidata	Store or retrieve application data
overobj	Get handle of object the pointer is over
popupstr	Get popup menu selection string
remapfig	Transform figure objects' positions
setptr	Set figure pointer
setstatus	Set status text string in figure
uiclearmode	Clears currently active interactive mode

Preferences	
addpref	Add preference
getpref	Get preference
rmpref	Remove preference
setpref	Set preference
ispref	Test for existence of preference

22.18 Character strings

help strfun

General	
char	Create character array (string)
strings	Help for strings
cellstr	Cell array of strings from char array
blanks	String of blanks
deblank	Remove trailing blanks

String tests	
iscellstr	True for cell array of strings
ischar	True for character array (string)
isspace	True for white space characters
isstrprop	Check category of string elements

Character set conversion	
native2unicode	Convert bytes to Unicode characters
unicode2native	Convert Unicode characters to bytes

String to number conversion	
num2str	Convert number to string
int2str	Convert integer to string
mat2str	Convert matrix to eval'able string
str2double	Convert string to double-precision value
str2num	Convert string matrix to numeric array
sprintf	Write formatted data to string
sscanf	Read string under format control

String operations

regexp	Match regular expression
regexpi	Match regular expression, ignoring case
regexprep	Replace string using regular expression
strcat	Concatenate strings
strvcat	Vertically concatenate strings
strcmp	Compare strings
strncmp	Compare first N characters of strings
strcmpi	Compare strings ignoring case
strncmpi	Compare first N characters, ignore case
strread	Read formatted data from string
strtrim	Remove insignificant whitespace
findstr	Find one string within another
strfind	Find one string within another
strjust	Justify character array
strmatch	Find possible matches for string
strrep	Replace string with another
strtok	Find token in string
upper	Convert string to uppercase
lower	Convert string to lowercase

Base number conversion

hex2num	IEEE hexadecimal to double-precision
hex2dec	hexadecimal string to decimal integer
dec2hex	decimal integer to hexadecimal string
bin2dec	Convert binary string to decimal integer
dec2bin	Convert decimal integer to binary string
base2dec	Convert base B string to decimal integer
dec2base	Convert decimal integer to base B string
num2hex	single and double to IEEE hexadecimal

22.19 Image and scientific data

```
help imagesci
```

Image file import/export

imformats	List details about supported file formats
imfinfo	Return information about graphics file
imread	Read image from graphics file
imwrite	Write image to graphics file
im2java	Convert image to Java image
multibandread	Read band-interleaved data from a file
multibandwrite	Write multiband data to a file

CDF file handling

cdfread	Read data from a CDF file
cdfinfo	Get information from a CDF file
cdfwrite	Write data to a CDF file
cdfepoch	Construct cdfepoch object

FITS file handling

fitsinfo	Get information from a FITS file
fitsread	Read data from a FITS file

HDF version 4 file handling

hdfinfo	Get information about an HDF4 file
hdfread	Read HDF4 file
hdftool	Browse/import HDF4 or HDF-EOS files

HDF version 5 file handling

hdf5info	Get information about an HDF5 file
hdf5read	Read data and attributes from HDF5 file
hdf5write	Write data and attributes to HDF5 file

HDF version 5 data objects

hdf5.h5array	Construct HDF5 array
hdf5.h5compound	Construct HDF5 compound object
hdf5.h5enum	Construct HDF5 enumeration object
hdf5.h5string	Construct HDF5 string
hdf5.h5vlen	Construct HDF5 variable length array

HDF version 4 library interface	
hdf	MEX-file interface to the HDF library
hdfan	HDF multifile annotation interface
hdfdf24	HDF raster image interface
hdfdfr8	HDF 8-bit raster image interface
hdfh	HDF H interface
hdfhe	HDF HE interface
hdfhx	HDF HX interface
hdfml	MATLAB-HDF gateway utilities
hdfsd	HDF multifile scientific dataset interface
hdfv	HDF V (Vgroup) interface
hdfvf	HDF VF (Vdata) interface
hdfvh	HDF VH (Vdata) interface
hdfvs	HDF VS (Vdata) interface

HDF-EOS library interface help	
hdfgd	HDF-EOS grid interface
hdfpt	HDF-EOS point interface
hdfsw	HDF-EOS swath interface

22.20 File input/output

help iofun

File import/export functions	
dlmread	Read ASCII delimited text file
dlmwrite	Write ASCII delimited text file
importdata	Load data from a file into MATLAB
daqread	Read Data Acquisition Toolbox daq file
matfinfo	Text description of MAT-file contents

Internet resource	
urlread	Read URL contents as a string
urlwrite	Save URL contents to a file
ftp	Create an ftp object
sendmail	Send e-mail

Spreadsheet support	
xlsread	Read Excel (xls) workbook
xlswrite	Write to Excel (xls) workbook
xlsfinfo	Check if file contains Excel workbook
wk1read	Read Lotus spreadsheet (wk1) file
wk1write	Write Lotus spreadsheet (wk1) file
wk1finfo	Check if file contains Lotus worksheet
str2rng	Convert range string to numeric array
wk1wrec	Write a Lotus worksheet record header

Zip file access	
zip	Compress files in a zip file
unzip	Extract contents of a zip file

Formatted file I/O	
fgetl	Read line from file, discard newline char
fgets	Read line from file, keep newline char.
fprintf	Write formatted data to file
fscanf	Read formatted data from text file
textscan	Read formatted data from text file
textread	Read formatted data from text file

File opening and closing	
fopen	Open file
fclose	Close file

Binary file I/O	
fread	Read binary data from file
fwrite	Write binary data to file

File positioning	
feof	Test for end-of-file
ferror	Inquire file error status
frewind	Rewind file
fseek	Set file position indicator
ftell	Get file position indicator

File name handling

fileparts	Filename parts
filesep	Directory separator for this platform
fullfile	Build full filename from parts
matlabroot	Root directory of MATLAB installation
mexext	MEX filename extension
partialpath	Partial pathnames
pathsep	Path separator for this platform
prefdir	Preference directory name
tempdir	Get temporary directory
tempname	Get temporary file

XML file handling

xmlread	Parse an XML document
xmlwrite	Serialize XML Document Object Model
xslt	Transform XML document via XSLT

Serial port support

serial	Construct serial port object
instrfindall	Find all serial port objects
freeserial	Release serial port
instrfind	Find serial port objects

Timer support

timer	Construct timer object
timerfindall	Find all timer objects
timerfind	Find visible timer objects

Command window I/O

clc	Clear Command window
home	Send cursor home

SOAP support

callSoapService	Send a SOAP message
createSoapMessage	Create a SOAP message
parseSoapResponse	Convert SOAP message response

22.21 Audio and video support

help audiovideo

Audio input/output objects

audioplayer	Audio player object
audiorecorder	Audio recorder object

Audio hardware drivers

sound	Play vector as sound
soundsc	Autoscale and play vector as sound
wavplay	Play on Windows audio output device
wavrecord	Record from Windows audio input

Audio file import and export

aufinfo	Return information about au file
auread	Read NeXT/SUN (.au) sound file
auwrite	Write NeXT/SUN (.au) sound file
wavfinfo	Return information about wav file
wavread	Read Microsoft (.wav) sound file
wavwrite	Write Microsoft (.wav) sound file

Video file import/export

aviread	Read movie (.avi) file
aviinfo	Return information about avi file
avifile	Create a new avi file
movie2avi	Make avi movie from MATLAB movie

Utilities

lin2mu	Convert linear signal to mu-law encoding
mu2lin	Convert mu-law encoding to linear signal

Example audio data (MAT files)

chirp	Frequency sweeps
gong	Gong
handel	Hallelujah chorus
laughter	Laughter from a crowd
splat	Chirp followed by a splat
train	Train whistle

22.22 Time and dates
`help timefun`

Current date and time	
now	Current date and time as date number
date	Current date as date string
clock	Current date and time as date vector

Basic functions	
datenum	Serial date number
datestr	String representation of date
datevec	Date components

Date functions	
calendar	Calendar
weekday	Day of week
eomday	End of month
datetick	Date formatted tick labels

Timing functions	
cputime	CPU time in seconds
tic	Start stopwatch timer
toc	Stop stopwatch timer
etime	Elapsed time
pause	Wait in seconds

22.23 Data types and structures
`help datatypes`

Class determination functions	
isnumeric	True for numeric arrays
isfloat	True for single and double arrays
isinteger	True for integer arrays
islogical	True for logical arrays
ischar	True for char arrays (string)

Data types (classes)

`double`	Convert to double precision
`char`	Create character array (string)
`logical`	Convert numeric values to logical
`cell`	Create cell array
`struct`	Create or convert to structure array
`single`	Convert to single precision
`int8`	Convert to signed 8-bit integer
`int16`	Convert to signed 16-bit integer
`int32`	Convert to signed 32-bit integer
`int64`	Convert to signed 64-bit integer
`uint8`	Convert to unsigned 8-bit integer
`uint16`	Convert to unsigned 16-bit integer
`uint32`	Convert to unsigned 32-bit integer
`uint64`	Convert to unsigned 64-bit integer
`inline`	Construct `inline` object
`function_handle`	Function handle (@ operator)
`javaArray`	Construct a Java array
`javaMethod`	Invoke a Java method
`javaObject`	Invoke a Java object constructor

Multidimensional array functions

`cat`	Concatenate arrays
`ndims`	Number of dimensions
`ndgrid`	Arrays for N-D functions & interpolation
`permute`	Permute array dimensions
`ipermute`	Inverse permute array dimensions
`shiftdim`	Shift dimensions
`squeeze`	Remove singleton dimensions

Function handle functions

`@`	Create `function_handle`
`func2str`	`function_handle` array to string
`str2func`	String to `function_handle` array
`functions`	List functions of a `function_handle`

Cell array functions

`cell`	Create cell array
`cellfun`	Functions on cell array contents
`celldisp`	Display cell array contents
`cellplot`	Display graphical depiction of cell array
`cell2mat`	Combine cell array of matrices
`mat2cell`	Break matrix into cell array of matrices
`num2cell`	Convert numeric array into cell array
`deal`	Deal inputs to outputs
`cell2struct`	Convert cell array into structure array
`struct2cell`	Convert structure array into cell array
`iscell`	True for cell array

Structure functions

`struct`	Create or convert to structure array
`fieldnames`	Get structure field names
`getfield`	Get structure field contents
`setfield`	Set structure field contents
`rmfield`	Remove structure field
`isfield`	True if field is in structure array
`isstruct`	True for structures
`orderfields`	Order fields of a structure array

Object-oriented programming functions

`class`	Create object or return object class
`struct`	Convert object to structure array
`methods`	List names & properties of class methods
`methodsview`	List names & properties of class methods
`isa`	True if object is a given class
`isjava`	True for Java objects
`isobject`	True for MATLAB objects
`inferiorto`	Inferior class relationship
`superiorto`	Superior class relationship
`substruct`	Create structure for `subsref`/`subasgn`

Overloadable operators	
`minus`	Overloadable method for `a-b`
`plus`	Overloadable method for `a+b`
`times`	Overloadable method for `a.*b`
`mtimes`	Overloadable method for `a*b`
`mldivide`	Overloadable method for `a\b`
`mrdivide`	Overloadable method for `a/b`
`rdivide`	Overloadable method for `a./b`
`ldivide`	Overloadable method for `a.\b`
`power`	Overloadable method for `a.^b`
`mpower`	Overloadable method for `a^b`
`uminus`	Overloadable method for `-a`
`uplus`	Overloadable method for `+a`
`horzcat`	Overloadable method for `[a b]`
`vertcat`	Overloadable method for `[a;b]`
`le`	Overloadable method for `a<=b`
`lt`	Overloadable method for `a<b`
`gt`	Overloadable method for `a>b`
`ge`	Overloadable method for `a>=b`
`eq`	Overloadable method for `a==b`
`ne`	Overloadable method for `a~=b`
`not`	Overloadable method for `~a`
`and`	Overloadable method for `a&b`
`or`	Overloadable method for `a|b`
`subsasgn`	for `a(i)=b`, `a{i}=b`, and `a.field=b`
`subsref`	for `a(i)`, `a{i}`, and `a.field`
`colon`	Overloadable method for `a:b`
`end`	Overloadable method for `a(end)`
`transpose`	Overloadable method for `a.'`
`ctranspose`	Overloadable method for `a'`
`subsindex`	Overloadable method for `x(a)`
`loadobj`	Called to load object from `.mat` file
`saveobj`	Called to save object to `.mat` file

22.24 Version control

`help verctrl`

Checkin/checkout	
checkin	checkin files to version control system
checkout	checkout files
undocheckout	undo checkout files

Specific version control	
rcs	Version control actions using RCS
pvcs	Version control actions using PVCS
clearcase	Version control actions using ClearCase
sourcesafe	Version control using Visual SourceSafe
customverctrl	Custom version control template
verctrl	Version control operations on Windows
cmpopts	Version control settings

22.25 Creating and debugging code

`help codetools`

Writing and managing M-files	
edit	Edit M-file
notebook	Open an M-book in Microsoft Word
mlint	List suspicious constructs in M-files

Directory tools	
contentsrpt	Audit Contents.m of a directory
coveragerpt	Scan directory for profiler line coverage
deprept	Scan file or directory for dependencies
diffrpt	Directory browser and file comparitor
dofixrpt	Scan file or directory for TODO, ...
helprpt	Scan file or directory for help
mlintrpt	Scan file or directory for M-lint info.
standardrpt	Directory browser

Managing the file system and search patch	
filebrowser	Open Current Directory browser
pathtool	View, modify, & save MATLAB path

Profiling M-files	
`profile`	Profile function execution time
`profview`	Profile browser
`profsave`	Save static copy of profile report
`profreport`	Generate profile report
`profviewgateway`	Profiler HTML gateway function
`opentoline`	Start editting a file at a given line
`stripanchors`	Remove code evaluation anchors

Debugging M-files	
`debug`	`help debug` lists debugging commands
`dbstop`	Set breakpoint
`dbclear`	Remove breakpoint
`dbcont`	Continue execution
`dbdown`	Change local workspace context
`dbstack`	Display function call stack
`dbstatus`	List all breakpoints
`dbstep`	Execute one or more lines
`dbtype`	List M-file with line numbers
`dbup`	Change local workspace context
`dbquit`	Quit debug mode
`dbmex`	Debug MEX-files (Unix only)

Managing, watching, and editting variables	
`openvar`	Open workspace for graphical editting
`workspace`	View contents of a workspace

22.26 Help commands
`help helptools`

Accessing on-line HTML help	
`doc`	Bring up Help Browser to specific place
`helpbrowser`	Same as doc, to last place viewed
`helpdesk`	Same as doc, to help "Begin Here" page
`helpview`	Display HTML file in Help Browser
`docsearch`	Search in Help Browser

Accessing M-file help	
help	View M-file help in Command window
helpwin	View M-file help in Help Browser
lookfor	Search all M-files for keyword

MathWorks tech support, web access	
info	Info about MATLAB & The MathWorks
support	Open MathWorks tech support web page
whatsnew	View Release Notes in Help Browser
web	Open internal or system web browser

22.27 Microsoft Windows functions

help winfun

COM automation client functions	
actxcontrol	Create an ActiveX control
actxserver	Create an ActiveX server
eventlisteners	Lists all registered events
isevent	True if event of object
registerevent	Registers events
unregisterallevents	Unregister all events
unregisterevent	Unregister events
iscom	True if COM/ActiveX object
isinterface	True if COM interface
COM/set	Set COM object property
COM/get	Get COM object properties
COM/invoke	Invoke/display method
COM/events	List COM object events
COM/interfaces	List custom interfaces
COM/addproperty	Add custom property to object
COM/deleteproperty	Remove custom property
COM/delete	Delete COM object
COM/release	Release COM interface
COM/move	Move/resize ActiveX control
COM/propedit	Edit properties

COM automation client functions (continued)	
`COM/save`	Serialize COM object to file
`COM/load`	Initialize COM object from file

COM sample code	
`mwsamp`	ActiveX control creation
`sampev`	Event handler

DDE client functions	
`ddeadv`	Setup advisory link
`ddeexec`	Execute string
`ddeinit`	Initiate DDE conversation
`ddepoke`	Send data to application
`ddereq`	Request data from application
`ddeterm`	Terminate DDE conversation
`ddeunadv`	Release advisory link

General MS Windows functions	
`winopen`	Open file using Windows command
`winqueryreg`	Read Windows registry

22.28 Examples and demonstrations

Type `help demos` to see a list of MATLAB demos.

22.29 Preferences
`help local`

Saved preferences files	
`startup`	User startup M-file
`finish`	User finish M-file
`matlabrc`	Master startup M-file
`pathdef`	Search path defaults
`docopt`	Web browser defaults
`printopt`	Printer defaults

Configuration information	
`hostid`	MATLAB server host ID number
`license`	License number
`version`	MATLAB version number

22.30 Symbolic Math Toolbox

`help symbolic`

Demonstrations	
`symintro`	Introduction to Symbolic Math Toolbox
`symcalcdemo`	Calculus demonstration
`symlindemo`	Demonstrate symbolic linear algebra
`symvpademo`	Variable precision arithmetic demo
`symrotdemo`	Study plane rotations
`symeqndemo`	Demonstrate symbolic equation solving

Symbolic operations	
`sym`	Create symbolic object
`syms`	Create symbolic object (short-hand)
`findsym`	Determine symbolic variables
`pretty`	Pretty print a symbolic expression
`latex`	Symbolic expression in LaTeX
`texlabel`	Convert string to TeX
`ccode`	Symbolic expression in C code
`fortran`	Symbolic expression in Fortran code

Calculus	
`diff`	Differentiate
`int`	Integrate
`limit`	Limit
`taylor`	Taylor series
`jacobian`	Jacobian matrix
`symsum`	Summation of series

Linear algebra

`diag`	Create or extract diagonals	
`triu`	Upper triangle	
`tril`	Lower triangle	
`inv`	Matrix inverse	
`det`	Determinant	
`rank`	Rank	
`rref`	Reduced row echelon form	
`null`	Basis for null space	
`colspace`	Basis for column space	
`eig`	Eigenvalues and eigenvectors	
`svd`	Singular values and singular vectors	
`jordan`	Jordan canonical (normal) form	
`poly`	Characteristic polynomial	
`expm`	Matrix exponential	
`mldivide`	Matrix left division (backslash)	`a\b`
`mpower`	Matrix power	`a^b`
`mrdivide`	Matrix right division (slash)	`a/b`
`mrtimes`	Matrix multiplication	`a*b`
`transpose`	Matrix transpose	`a.'`
`ctranspose`	Matrix complex conj. transpose	`a'`

Simplification

`simplify`	Simplify
`expand`	Expand
`factor`	Factor
`collect`	Collect
`simple`	Search for shortest form
`numden`	Numerator and denominator
`horner`	Nested polynomial representation
`subexpr`	Rewrite in terms of subexpressions
`coeffs`	Coefficients of a multivariate polynomial
`sort`	Sort symbolic vectors or polynomials
`subs`	Symbolic substitution

Solution of equations	
solve	Solve algebraic (nonlinear) equations
dsolve	Solve differential equations
finverse	Functional inverse
compose	Functional composition

Variable precision arithmetic	
vpa	Variable precision arithmetic
digits	Set variable precision accuracy

Integral transforms	
fourier	Fourier transform
laplace	Laplace transform
ztrans	Z transform
ifourier	Inverse Fourier transform
ilaplace	Inverse Laplace transform
iztrans	Inverse Z transform

Conversions	
double	Convert symbolic matrix to double
single	Convert symbolic matrix to single
poly2sym	Coefficients to symbolic polynomial
sym2poly	Symbolic polynomial to coefficients
char	Convert sym object to string
int8	Convert to signed 8-bit integer
int16	Convert to signed 16-bit integer
int32	Convert to signed 32-bit integer
int64	Convert to signed 64-bit integer
uint8	Convert to unsigned 8-bit integer
uint16	Convert to unsigned 16-bit integer
uint32	Convert to unsigned 32-bit integer
uint64	Convert to unsigned 64-bit integer

Arithmetic and algebraic operations

plus	Addition	a+b
minus	Subtraction	a-b
uminus	Negation	-a
times	Array multiplications	a.*b
ldivide	Left division (backslash)	a\b
rdivide	Right division (slash)	a/b
power	Array power	a.^b
abs	Absolute value	
ceil	Ceiling	
conj	Conjugate	
colon	Colon operator	
fix	Integer part	
floor	Floor	
frac	Fractional part	
mod	Modulus	
round	Round	
quorem	Quotient and remainder	
imag	Imaginary part	
real	Real part	
exp	Exponential	
log	Natural logarithm	
log10	Common (base-10) logarithm	
log2	Base-2 logarithm	
sqrt	Square root	
prod	Product of elements	
sum	Sum of elements	

Logical operations

isreal	True for real array	
eq	Equality test a==b	
ne	Inequality test a~=b	

Trigonometric functions

sin	Sine
sinh	Hyperbolic sine
asin	Inverse sine
asinh	Inverse hyperbolic sine
cos	Cosine
cosh	Hyperbolic cosine
acos	Inverse cosine
acosh	Inverse hyperbolic cosine
tan	Tangent
tanh	Hyperbolic tangent
atan	Inverse tangent
atanh	Inverse hyperbolic tangent
sec	Secant
sech	Hyperbolic secant
asec	Inverse secant
asech	Inverse hyperbolic secant
csc	Cosecant
csch	Hyperbolic cosecant
acsc	Inverse cosecant
acsch	Inverse hyperbolic cosecant
cot	Cotangent
coth	Hyperbolic cotangent
acot	Inverse cotangent
acoth	Inverse hyperbolic cotangent

String handling utilities

isvarname	Check for a valid variable name
vectorize	Vectorize a symbolic expression
disp	Display symbolic expression as text
display	Display function for symbolic statements
eval	Evaluate a symbolic expression

Special functions	
`besselj`	Bessel function of the first kind
`bessely`	Bessel function of the second kind
`besseli`	Modified Bessel function of the 1st kind
`besselk`	Modified Bessel function of the 2nd kind
`erf`	Error function
`sinint`	Sine integral
`cosint`	Cosine integral
`zeta`	Riemann zeta function
`gamma`	Symbolic gamma function
`gcd`	Greatest common divisor
`lcm`	Least common multiple
`hypergeom`	Generalized hypergeometric function
`lambertw`	Lambert W function
`dirac`	Delta function
`heaviside`	Step function

Pedagogical and graphical applications	
`rsums`	Riemann sums
`ezcontour`	Easy-to-use contour plotter
`ezcontourf`	Easy-to-use filled contour plotter
`ezmesh`	Easy-to-use mesh (surface) plotter
`ezmeshc`	Easy-to-use mesh/contour plotter
`ezplot`	Easy-to-use function plotter
`ezplot3`	Easy-to-use spatial curve plotter
`ezpolar`	Easy-to-use polar coordinates plotter
`ezsurf`	Easy-to-use surface plotter
`ezsurfc`	Easy-to-use surface/contour plotter
`funtool`	Function calculator
`taylortool`	Taylor series calculator

Access to Maple (not in Student Version)	
`maple`	Access Maple kernel
`mfun`	Numeric evaluation of Maple functions
`mfunlist`	List of functions for `mfun`
`mhelp`	Maple help

23. Additional Resources

The MathWorks, Inc., and others provide a wide range of products that extend MATLAB's capabilities. Some are collections of M-files called toolboxes. One of these has already been introduced (the Symbolic Math Toolbox). Also available is Simulink, an interactive graphical system for modeling and simulating dynamic nonlinear systems. The `ver` command lists the toolboxes and Simulink components included in your installation, as does the Help Browser (`doc`). Similar to MATLAB toolboxes, Simulink has domain-specific add-ons called blocksets.

MATLAB:

MATLAB®
Database Toolbox
MATLAB Report Generator

Math and Optimization:
Optimization Toolbox
Symbolic Toolbox
Extended Symbolic Math Toolbox
Partial Differential Equation Toolbox
Genetic Algorithm and Direct Search Toolbox

Statistics and Data Analysis:
Statistics Toolbox
Neural Network Toolbox
Curve Fitting Toolbox
Spline Toolbox
Model-Based Calibration Toolbox
Bioinformatics Toolbox

Control System Design and Analysis:
Control System Toolbox
System Identification Toolbox
Fuzzy Logic Toolbox
Robust Control Toolbox
μ-Analysis and Synthesis Toolbox
LMI Control Toolbox
Model Predictive Control Toolbox

Signal Process and Communications:
Signal Processing Toolbox
Communications Toolbox
Filter Design Toolbox
Filter Design HDL Coder
System Identification Toolbox
Wavelet Toolbox
Fixed-Point Toolbox
RF Toolbox
Link for Code Composer Studio™
Link for ModelSim®

Image Processing:
Image Processing Toolbox
Image Acquisition Toolbox
Mapping Toolbox

Test and Measurement:
Data Acquisition Toolbox
Instrument Control Toolbox
Image Acquisition Toolbox
OPC Toolbox

Financial Modeling and Analysis:
Financial Toolbox

Financial Derivatives Toolbox
GARCH Toolbox
Financial Time Series Toolbox
Datafeed Toolbox
Fixed-Income Toolbox

Application Deployment:
MATLAB Compiler
Excel Link
MATLAB Web Server

Application Deployment Targets:
MATLAB Builder for COM
MATLAB Builder for Excel

Simulink:

Simulink®
Stateflow®
Simulink Fixed Point
Simulink Accelerator
Simulink Report Generator

Physical Modeling:
SimMechanics
SimPowerSystems

Simulation Graphics:
Virtual Reality Toolbox
Dials and Gauges Blockset

Control System Design and Analysis:
Simulink Control Design
Simulink Response Optimization

Simulink Parameter Estimation
Aerospace Blockset

Signal Processing and Communications:
Signal Processing Blockset
Communications Blockset
CDMA Reference Blockset
RF Blockset

Code Generation:
Real-Time Workshop®
Real-Time Workshop Embedded Coder
Stateflow Coder

PC-Based Rapid Control Prototyping and HIL:
xPC Target
xPC Target Embedded Option
xPC TargetBox™
Real-Time Windows Target

Embedded Targets:
Embedded Target for TI C6000™ DSP
Embedded Target for Motorola® MPC555
Embedded Target for OSEK/VDX®
Embedded Target for Infineon C166®
    Microcontrollers
Embedded Target for Motorola® HC12
Embedded Target for TI C2000™ DSP

Verification, Validation, and Testing:
Link for Code Composer Studio™
Link for ModelSim®
Simulink Verification and Validation

Index

script, 35, 36, 149
Microsoft
  Excel, 181, 200
  Powerpoint, 132
  Windows, vi, 1, 2,
    54, 145, 183, 190
  Word, 132
min, 24, 158
mod, 32, 154, 195
more, 7, 144
multidimensional
  arrays, 17, 185
multiple output. *See* [].
mxArray, 54
mxCalloc, 59
mxCopyPtrToReal8,
  64
mxCopyReal8ToPtr,
  64
mxCreateDoubleMatrix,
  56, 59, 62
mxFree, 56, 59
mxGetM, 56, 59
mxGetN, 56, 59, 62
mxGetPr, 56, 59, 62
mxGetScalar, 56, 59
mxIsEmpty, 56, 59
mxIsSparse, 56, 59
mxMalloc, 56, 59
name resolution, 47
nargin, 42, 46, 149
nargout, 42, 46, 149
nnz, 85, 163

nonlinear equations. *See*
  equations.
norm, 20, 25, 50, 126,
  156
normest, 25, 156, 164
not-a-number, 150, 151
null, 110, 156, 193
null space. *See* null.
num2str, 131, 177
ode45, 125, 162
ones, 21, 87, 150
ones matrix. *See* ones,
  spones.
open icon, 78
openvar, 9, 189
operating system, 67,
  69, 145
operators
  arithmetic, 11, 146,
    187, 195
  bit-wise, 147
  entry-wise, 13, 23,
    44, 71, 79, 80, 110,
    146, 187, 195
  index, 10, 19, 147
  logical, 14, 146, 187,
    195
  matrix, 11, 146, 156,
    187, 193, 195
  overloadable, 187
  relational, 13, 27, 91,
    146, 187, 195
  scalar, 11
  set, 148